The physics of vibration

Volume 2

The physics of vibration

Volume 2, containing Part 2, The simple vibrator in quantum mechanics

A. B. PIPPARD, FRS

Cavendish Professor of Physics, University of Cambridge

CAMBRIDGE UNIVERSITY PRESS

Cambridge
London New York New Rochelle
Melbourne Sydney

Published by the Press Syndicate of the University of Cambridge
The Pitt Building, Trumpington Street, Cambridge CB2 1RP
32 East 57th Street, New York, NY 10022, USA
296 Beaconsfield Parade, Middle Park, Melbourne 3206, Australia

First published 1983

Printed in Great Britain at the University Press, Cambridge

Library of Congress catalogue card number: 77–85685

British Library Cataloguing in Publication Data

Pippard, A. B.
The physics of vibration.
Vol. 2
1. Vibration
I. Title
531′.32 QC136

ISBN 0 521 24623 7

Contents

Note: Equations, diagrams and references are numbered serially in each chapter, and the chapter number is omitted if it is something in the same chapter that is referred to. Thus (14.19) in chapter 18 means equation 19 of chapter 14; (19) in chapter 18 means equation 19 in that chapter.

Cross-references are given as page-numbers in the margin; I.127 means page 127 of Volume 1; 127 means page 127 of this volume.

Preface

Ten years ago, when I started writing on the physics of vibration, I had in mind a single volume. Four years ago I was reconciled to the need for two, and now must confess that the complexity of the subject has made it advisable to get the second of the projected three parts into print without waiting for the third to accompany it between the same covers. It is still my hope to do justice to the vibrations of extended systems, but the difficulties are considerable and not made easier by the vigour with which some of the central topics are being pursued at present.

Of all the encouragement I have enjoyed I particularly wish to record with the warmest thanks the help of Dr Edmund Crouch and Dr John Hannay who long ago, as research students, derived for me some solutions of Schrödinger's equation which provided a stout anchor for my thoughts: Dr Bob Butcher who has firmly guided me in my brief excursions into structural chemistry and molecular spectra: and Dr Andrew Phillips who devoted more time than he could have been expected to spare to a critical reading of much of the typescript. If the state of that typescript as delivered to the printers did not achieve even a modest standard of tidiness, the fault is entirely the consequence of copious afterthoughts on my part, and in no way to be blamed on Mrs Janet Thulborn whose patient and faithful typing deserved better respect, and has certainly earned my gratitude. I am happy to acknowledge the generous provision by Dr Brian Petley of the original of fig. 19.5. The typography of the Cambridge University Press continues to give me great satisfaction, but an oversight of mine in volume 1 has been criticized – without chapter numbers at the head of each page it is unnecessarily difficult to locate an equation. This fault has been avoided in the present volume.

A. B. PIPPARD

Cambridge 1981

Part 2

The simple vibrator in quantum mechanics

13 The quantized harmonic vibrator and its classical features

By the time a student of physics is ready to tackle quantum mechanics he has become familiar with the classical harmonic vibrator through many examples, and knows the crucial role it played in the development of Planck's ideas. It is natural then to concentrate on the mathematical aspects of the harmonic oscillator equation in quantum mechanics, the solution of Schrödinger's equation, normalization of the wave-function, calculation of mean values and of matrix elements, leaving the physics to look after itself. At a more advanced level the harmonic vibrator provides the entry into field theories, second quantization etc., and as a general rule it tends to be viewed more as a vehicle for instruction in more interesting matters than as a physical system having its own considerable interest and importance. Here we shall seek to redress the balance and study the vibrator as a thing in itself, without losing sight of the variety of physical problems to which the results can be applied. One especially important set of applications, however, will get little attention at this stage – the vibrations of compound bodies and of extended physical systems provide such wealth of interest as to justify a volume to themselves; and it is to such matters that the whole of the third part of this work will be devoted. As soon as one begins to contemplate the harmonic vibrator one becomes aware of the exceptional nature of its behaviour, in that it conforms more closely than any other system to the classical rules. It is surprising how many problems involving harmonic vibrators are found to have the same solution in classical and quantum mechanics, and we shall pay considerable attention to this question with a view to defining as clearly as possible the limits of validity of the classical approach.

There is always a danger in a venture like this of being misunderstood as attempting to obviate the need for quantum mechanics. Undoubtedly life would be simpler if the classical solution to every problem were correct; we need look no further than the next chapter for examples of non-linear vibrators which yield to classical methods without trouble but are distinctly less tractable in their quantized guise. But the classical solutions are not quite right, or sometimes are entirely wrong, so that however much we may value the insight that classical methods provide they can never be regarded as valid alternatives. Indeed, we have only to glance back once more to Planck to recognize that it was the non-classical features of that most nearly classical system, the harmonic vibrator, that started the whole quantum revolution on its course. Nevertheless there is another side to the coin; if it had been necessary at every stage to solve Schrödinger's equation before being able to understand in qualitative terms the transport of electricity in metals and semiconductors, solid state physics would not be the highly developed science it is nowadays. It was the insight provided by classical visualization of the processes which gave a firm structure to detailed quantum-mechanical analyses; and it is to make available these classical insights, and to delineate their range of validity, that we undertake this study of classical correspondences. All the same, to allay any residual fears of a reactionary attack on modern physics, we shall begin with an outline of the conventional approach to the quantized oscillator through the formal solution of Schrödinger's equation. No more than an outline is needed since the details are available in countless textbooks of quantum mechanics.[(1),(2)] We shall then have available the elementary results needed for the next stages of the analysis. It is worth remarking at this point that the treatment throughout will be elementary, based on the direct solution of Schrödinger's equation. This is not a textbook of quantum mechanics, but a study of real physical vibrators, and there is no call to employ sophisticated mathematical procedures that were developed either for the sheer love of elegance or for their direct descriptive power when applied to more deeply quantal problems having little or no point of contact with classical ideas. Even so, it is possible that the unconventionally thorough discussion of elementary matters will also enlighten the reader who is familiar with advanced methods.

One example of the general attitude to be adopted is provided by the theory of dissipative processes, such as a charged harmonic vibrator radiating electromagnetic energy. We shall show how the classical theory of coupled oscillators may be applied to this problem in such a way as to give confidence in the interpretation of processes which are commonly presented as rather mysterious, but essentially quantal: spontaneous and stimulated transitions.

In so far as the emphasis in this chapter leans towards classical processes, it is to indicate the value of classical reasoning where it yields the correct answers. The following three chapters also draw on classical methods, but with a rather different end in view. There are a few standard problems – the harmonic oscillator, the hydrogen atom, the square well – for which complete analytical solution of Schrödinger's equation is feasible and useful. There are many more whose analysis involves unfamiliar series expansions or other difficulties, such that the labour of solution seems disproportionate to the desired end. For these, and anharmonic vibrators come into this category, it is often possible to derive rather good approximations to the correct solutions by the semi-classical approach associated with the names 32 of Bohr, Wilson and Sommerfeld; this is essentially a classical procedure supplemented by quantization rules which define the acceptable members of the continuous set of classical solutions. Between this approach and the solution of Schrödinger's equation lies the approximate method of Wentzel, 37

Kramers, Brillouin and Jeffreys. The next chapter illustrates the application of these techniques to a number of anharmonic vibrators which can also be treated exactly without too much difficulty. The power of the approximations, which is considerable, is thus revealed explicitly.

A third application of classical methods is developed in chapter 17; this is the replacement of a real physical system by a quite different classical system whose behaviour models certain features of the quantal behaviour of the real system. In particular the impulse response function of a quantized system, for instance an atom, can be modelled in the linear approximation by a set of classical harmonic oscillators. Consequently one may derive the response of the system to any weak time-dependent force from one's knowledge of the response of a harmonic oscillator. The possibility of designing such a model accounts for the success of classical interpretations of anomalous dispersion, and a similar modelling justifies the classical treatment of nuclear magnetic resonance.

86

The remainder of the volume builds on this identification of quantum transitions with classical harmonic oscillators, always with an eye to those features that have relevance to real vibrating physical systems. Thus the last two chapters deal with masers, lasers (but only very briefly) and bunching processes which form a conceptual link between masers and certain maintained oscillators, of which the klystron is selected for more detailed discussion. To reach this point it is necessary to examine the interaction of radiation with quantized systems in a number of different ways until, in chapter 20, there emerges a rather precise picture of a radiation-damped transition which, apart from a strictly quantal noise source, corresponds remarkably closely with the primitive classical model of a viscous-damped harmonic vibrator.

164

At various points in this development we have to face the question – if classical methods work so well for so many aspects of the behaviour of harmonic vibrators, wherein lies the essential quantal character? The answer we shall give is that the classical approach, however successful in describing a vibrator acted upon by a well-defined force, fails when the force is itself provided by a system that must be treated quantally. To understand this point is to possess something like an operational criterion for deciding whether classical methods are safe.

Solution of Schrödinger's equation

A particle of mass m, moving on a line under the influence of a linear restoring force, has potential energy V equal to $\frac{1}{2}m\omega_0^2x^2$ if its classical angular frequency of vibration is ω_0, and is governed by the time-independent Schrödinger equation:

$$(\hbar^2/2m)\,\mathrm{d}^2\psi/\mathrm{d}x^2+(E-\tfrac{1}{2}m\omega_0^2x^2)\psi=0. \tag{1}$$

In almost everything that follows we shall use reduced variables, setting

$\xi=(m\omega_0/\hbar)^{1/2}x$ and $\varepsilon=2E/\hbar\omega_0$, so that (1) takes the form:

$$\mathrm{d}^2\psi/\mathrm{d}\xi^2+(\varepsilon-\xi^2)\psi=0. \tag{2}$$

Note that the energy quantum, $\hbar\omega_0$, corresponds to $\varepsilon=2$ and the unit of length, $(\hbar/m\omega_0)^{1/2}$, is the amplitude of the corresponding classical oscillator when its energy is unity, or $\frac{1}{2}\hbar\omega_0$ in laboratory units.

One may choose ε at will and, having also chosen values of ψ and ψ' at some ξ, proceed to integrate (2) step by step. The resulting solution will almost certainly diverge rapidly at both positive and negative ξ, and hence cannot be regarded as physically meaningful. Whether it diverges or not, however, this solution can be used to generate another equally good solution for the same value of ε, simply by reflecting it in the origin; for the occurrence of ξ in (2) only in quadratic form shows that if $\psi(\xi)$ is a solution, so also is $\psi(-\xi)$.

Since (2) is a linear equation, all solutions may be cast in symmetric or antisymmetric form by taking $\psi(\xi)\pm\psi(-\xi)$ as the standard pattern. Instead of starting the integration with arbitrary choices of ψ and ψ' we may ensure one or other of those forms by starting always at the origin, and either taking $\psi'(0)$ as zero and $\psi(0)$ as non-zero for a symmetric solution, or $\psi'(0)$ as non-zero and $\psi(0)$ as zero for an antisymmetric solution. For every choice of ε there is one symmetric and one antisymmetric solution, and in general both diverge at large $|\xi|$. For when $\xi^2>\varepsilon$ the sign of ψ'' is the same as that of ψ, and from this point on divergence is almost inevitable. For if, as integration proceeds, we find ψ' taking the same sign as ψ, the curve can only move steadily away from the axis, since the sign of ψ' can never be reversed. If, on the other hand, the curve is pointing towards the axis, with ψ and ψ' opposite in sign, the slope may not decrease fast enough to prevent the curve crossing the axis; as soon as this happens divergence is inevitable, for ψ and ψ' have now the same sign. Convergence to zero at high values of ξ can only be achieved by the correct choice of ε so that the curve comes down to the axis at a grazing angle.

When ξ is very large the value of ε is almost irrelevant and by selecting convenient values the asymptotic behaviour can be readily determined. Thus when $\varepsilon=1$ one of the two independent solutions of (2) is $\mathrm{e}^{-\frac{1}{2}\xi^2}$, and when $\varepsilon=-1$ one of the solutions is $\mathrm{e}^{+\frac{1}{2}\xi^2}$. Both are very nearly solutions for other values of ε, and as ξ increases the solution for any ε approaches a linear combination of the two:

$$\psi\sim A\,\mathrm{e}^{\frac{1}{2}\xi^2}+B\,\mathrm{e}^{-\frac{1}{2}\xi^2}.$$

Only by choosing ε correctly can the symmetric or antisymmetric solution for small ξ join smoothly to the asymptotic form containing B only, and therefore converging to zero as required. This is the background for proceeding to discover solutions of (2) in the form of $\mathrm{e}^{-\frac{1}{2}\xi^2}$ multiplying a power series in ξ. The details are well presented in most standard textbooks, and we need only note here that the resulting power series in general diverges roughly as e^{ξ^2}, showing that in general $A\neq0$ in the asymptotic form. By choosing ε correctly, however, the series may be terminated so that ψ is $\mathrm{e}^{-\frac{1}{2}\xi^2}$ times a finite polynomial, $H_n(\xi)$, which can never increase so rapidly at large ξ as to overwhelm the exponential factor. The required values of ε are $2n+1$, corresponding to $E=(n+\frac{1}{2})\hbar\omega_0$ as conjectured by Planck except for the extra $\frac{1}{2}\hbar\omega_0$. We have then as acceptable solutions

$$\psi_n = \mathcal{N}_n H_n(\xi)\,\mathrm{e}^{-\frac{1}{2}\xi^2}, \tag{3}$$

in which the $\mathcal{N}_n$ are normalizing coefficients to ensure that $\int_{-\infty}^{\infty} \psi_n^2\,\mathrm{d}\xi = 1$:

$$\mathcal{N}_n = \pi^{-\frac{1}{4}}(2^n n!)^{-\frac{1}{2}}. \tag{4}$$

The $H_n(\xi)$ are Hermite polynomials, shown bracketed in (5), which is a sample of some of the lower-order wave-functions.[3]

$$\left.\begin{aligned}
n=0 \quad & \psi_0 = 0.75113\,(1)\,\mathrm{e}^{-\frac{1}{2}\xi^2}\\
1 \quad & \psi_1 = 0.53113\,(2\xi)\,\mathrm{e}^{-\frac{1}{2}\xi^2}\\
2 \quad & \psi_2 = 0.26556\,(4\xi^2-2)\,\mathrm{e}^{-\frac{1}{2}\xi^2}\\
3 \quad & \psi_3 = 0.10842\,(8\xi^3-12\xi)\,\mathrm{e}^{-\frac{1}{2}\xi^2}\\
\vdots \quad & \\
10 \quad & \psi_{10} = 1.2321\times 10^{-5}(1024\xi^{10}-23\,040\xi^8\\
& \quad +161\,280\xi^6-403\,200\xi^4+302\,400\xi^2-302\,40)\,\mathrm{e}^{-\frac{1}{2}\xi^2}
\end{aligned}\right\} \tag{5}$$

The Hermite polynomials are related by the equation

$$H_{n+1} = 2\xi H_m - 2nH_{n-1}, \tag{6}$$

so that it is easy to generate successive polynomials and hence as long a sequence of ψ_n as may be required. The examples given show the alternation of symmetric wave-functions (even powers of ξ) and antisymmetric (odd powers).

The form of a few wave-functions is shown in fig. 1, together with $\psi\psi^*$; since ψ is normalized, $\psi\psi^*\,\mathrm{d}\xi$ is the probability of finding the oscillating particle in the range $\mathrm{d}\xi$. For a classical oscillator this is the function $\mathrm{d}\xi/[\pi(\xi_0^2-\xi^2)^{\frac{1}{2}}]$, strongly peaked at the extremities, $\pm\xi_0$, and the envelope of $\psi\psi^*$ approximates to this form when n is large. In the lowest states, however, there is no sign of this, and in particular the ground state probability distribution retains a certain width, comparable to that of a classical oscillator with energy $\frac{1}{2}\hbar\omega_0$, the zero-point energy of the quantized system.

When n is large the extreme peak becomes sharper, as can be seen by examining the form of (2) in the vicinity of the classical limit, $\xi^2=\varepsilon$. Substituting $\varepsilon^{\frac{1}{2}}+z/2^{\frac{1}{3}}\varepsilon^{\frac{1}{6}}$ for ξ, we find (2) transformed into Airy's equation:

$$\mathrm{d}^2\psi/\mathrm{d}z^2 - z\psi = 0, \tag{7}$$

if a second-order term in z is neglected. This is equivalent to treating the parabolic curve for V as a straight line in the immediate neighbourhood of the classical limit. The dimensionless form of (7) indicates that the

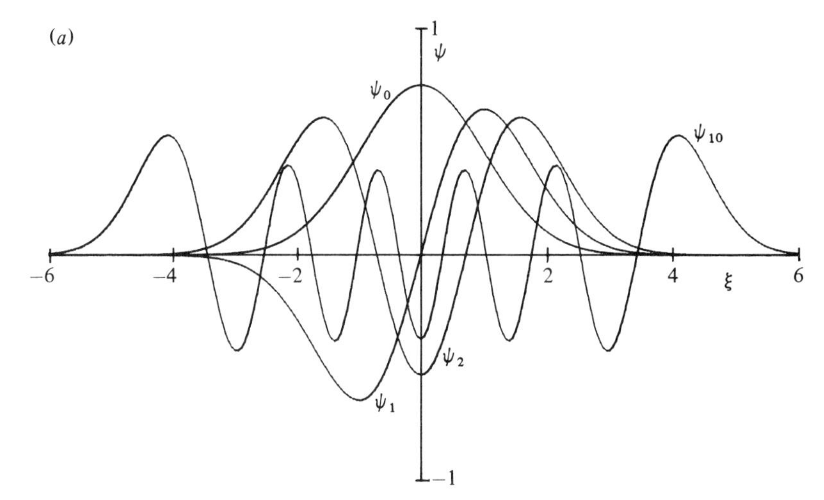

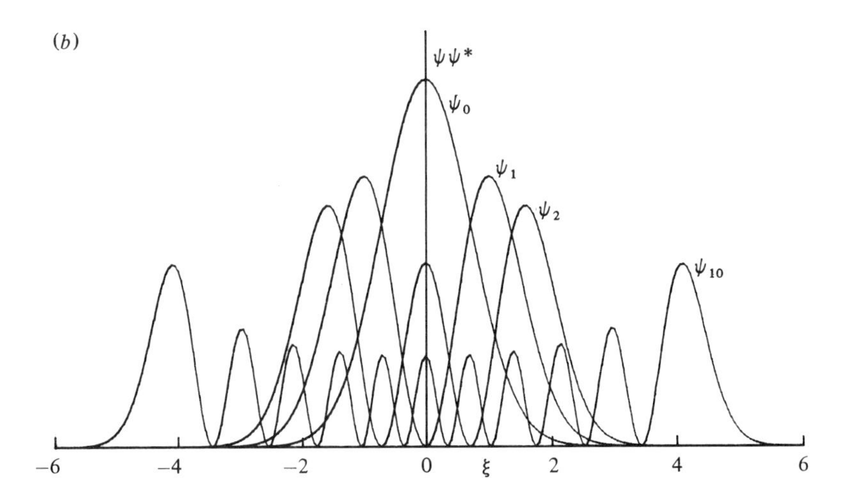

Fig. 1. (*a*) Wavefunctions of the harmonic vibrator stationary states ψ_0, ψ_1, ψ_2 and ψ_{10}, as given in (5). (*b*) $\psi\psi^*$ for the same wave-functions.

separation of the last two peaks is always the same when z is the co-ordinate, and that therefore in terms of ξ or the laboratory co-ordinate x the separation varies as $\varepsilon^{-\frac{1}{6}}$ – not a strong dependence but significant when the quantum number is 10^6 or more, such as we shall find when we come to discuss masers. By the same argument it follows that the tail of the wave-function, beyond the classical limit, narrows with increasing n. From a table of Airy functions[4] we find that the last peak of $\psi\psi^*$ occurs at $z = -1.019$ and that the value of $\psi\psi^*$ has dropped by a factor e just outside the classical limit at $z = 0.116$. Taking $\Delta z \times 1.135$ as a measure of the width of the tail, we have that corresponding $\Delta\xi_n = 0.80/n^{\frac{1}{6}}$, or 0.08 when $n = 10^6$. This may be compared with the width $\Delta\xi_0 = 1$ for the decay of $\psi_0\psi_0^*$ by a factor e from its peak

Momentum distribution

An alternative representation of the wave-function of a stationary state is as the superposition of travelling waves:

$$\psi_n(\xi) = \int_{-\infty}^{\infty} \varphi_n(\kappa)\, e^{i\kappa\xi}\, d\kappa. \tag{8}$$

Then $\varphi_n\varphi_n^*\, d\kappa$ represents the probability of finding the particle moving with wavenumber in the range $d\kappa$; and since $p \propto \kappa$, $\varphi_n\varphi_n^*$ describes the probability distribution for the momentum, p. By the Fourier transform theorem,

$$\varphi_n(\kappa) = \frac{1}{2\pi}\int_{-\infty}^{\infty} \psi_n(\xi)\, e^{-i\kappa\xi}\, d\xi. \tag{9}$$

We now demonstrate one of the peculiarities of the harmonic oscillator, that the differential equation obeyed by φ_n has the same form as that obeyed by ψ_n. We multiply each term in (2) by $e^{-i\kappa\xi}$ and integrate over ξ, making use of the following results obtained by differentiating within the integral signs of (8) and (9):

$$d^2\psi_n/d\xi^2 = -\int_{-\infty}^{\infty} \kappa^2\varphi_n\, e^{i\kappa\xi}\, d\kappa,$$

so that
$$\int_{-\infty}^{\infty} e^{-i\kappa\xi}(d^2\psi_n/d\xi^2)\, d\xi = -2\pi\kappa^2\varphi_n;$$

and
$$\int_{-\infty}^{\infty} \xi^2\psi_n\, e^{-i\kappa\xi}\, d\xi = -2\pi\, d^2\varphi_n/d\kappa^2.$$

Hence, from (2),

$$d^2\varphi_n/d\kappa^2 + (\varepsilon_n - \kappa^2)\varphi_n = 0. \tag{10}$$

The fact that this is the same as (2) shows that the diagrams in fig. 1 equally well represent the momentum distribution, just as with the classical vibrator, for which the peaked distribution is equally good for position and momentum, if the co-ordinates are suitably scaled. Our present choice of variables has this property. It may be noted that, since obviously $\langle\xi^2\rangle = \langle\kappa^2\rangle$, the mean values of kinetic and potential energy are the same. For just as $\xi = (m\omega_0/\hbar)^{\frac{1}{2}}x$, so $\kappa = (\hbar/m\omega_0)^{\frac{1}{2}}k$, k being the wavenumber measured in ordinary units, and

$$\left.\begin{aligned} \langle T\rangle &= (\hbar^2/2m)\langle k^2\rangle = \tfrac{1}{2}\hbar\omega_0\langle\kappa^2\rangle, \\ \text{while}\quad \langle V\rangle &= \tfrac{1}{2}m\omega_0^2\langle x^2\rangle = \tfrac{1}{2}\hbar\omega_0\langle\xi^2\rangle. \end{aligned}\right\} \tag{11}$$

This equality of mean kinetic and potential energy is also a classical property, and we shall show presently that it remains true even when the oscillator occupies a non-stationary mixture of states. For this purpose, however, we must consider the representation of non-stationary states by the superposition of stationary states.

Non-stationary states

The stationary state wave-functions, ψ_n, are the spatial parts of solutions of Schrödinger's time-dependent wave equation which, in one dimension, takes the form:

$$(\hbar^2/2m)\partial^2\Psi/\partial x^2 - V\Psi = -i\hbar\partial\Psi/\partial t. \tag{12}$$

When $V = \frac{1}{2}m\omega_0^2x^2$ the appropriate form for the harmonic vibrator in reduced variables may be written

$$\partial^2\Psi/\partial\xi^2 - \xi^2\Psi = -2i\partial\Psi/\partial\tau, \tag{13}$$

in which $\xi = (m\omega_0/\hbar)^{\frac{1}{2}}x$ as before, and $\tau = \omega_0 t$. The full expression for the wave-function Ψ_n of the nth stationary state is $\psi_n\, \mathrm{e}^{-\mathrm{i}(n+\frac{1}{2})\tau}$ where ψ_n satisfies (2) if $\varepsilon = 2n+1$. Since (13) is linear in Ψ, arbitrary superpositions of various Ψ_n are also solutions:

$$\Psi = \sum_n a_n\Psi_n = \mathrm{e}^{-\frac{1}{2}i\tau}\sum_n a_n\psi_n\,\mathrm{e}^{-in\tau}. \tag{14}$$

When Ψ and all ψ_n are normalized to unity, as is customary,

$$1 = \int_{-\infty}^{\infty}\Psi\Psi^*\,\mathrm{d}\xi = \sum_n\sum_m a_n a_m^*\int_{-\infty}^{\infty}\psi_n\psi_m^*\,\mathrm{d}\xi = \sum_n a_n a_n^*, \tag{15}$$

since the ψ_n are orthogonal. One may interpret $a_n a_n^*$ as the probability of finding the system in the nth state, in an experiment designed to yield such information.† Or if one seeks to determine the energy of the system the mean value $\langle E\rangle$ for a large number of repetitions of the measurement is $\Sigma\, a_n a_n^* E_n$, the weighted average of the energy levels over the occupation probability.

So long as the system is left undisturbed, with V independent of time, Ψ keeps the form (14) with all a_n constant. Once the form of Ψ at any moment has been expressed in terms of a sum of ψ_n, the subsequent development of Ψ follows inexorably. Moreover the ψ_n obtained by finding all stationary solutions of Schrödinger's time-independent equation, with any chosen variation of V with position, constitute a complete set of orthogonal functions, and may be used to express as a Fourier sum any other function whose behaviour is not too pathological. If, therefore, Ψ is known at any instant, the recipe for determining its subsequent development is to express it as $\Sigma_n\, a_n\psi_n$ and then to supplement each ψ_n with its appropriate time-variation, as in (14).

† The Stern–Gerlach experiment is such an experiment, atoms in different states being spatially segregated by an inhomogeneous magnetic field; it is not easy to devise a corresponding experiment for a harmonic vibrator.

The coherent state

As an example consider a function which can easily be verified as a normalized solution of (13):

$$\Psi = \pi^{-\frac{1}{4}} \mathrm{e}^{-\frac{1}{2}(\xi - b\cos\tau)^2 - \mathrm{i}\theta(\xi,\tau)} \tag{16}$$

in which b is a real constant and

$$\theta = \tfrac{1}{2}\tau + b\xi \sin\tau - \tfrac{1}{4}b^2 \sin 2\tau. \tag{17}$$

The phase θ disappears, of course, from the probability distribution:

$$\mathscr{P}(\xi) = \Psi\Psi^* = \pi^{-\frac{1}{2}} \mathrm{e}^{-(\xi - b\cos\tau)^2}. \tag{18}$$

The wave-function (16) is a Gaussian wave-packet of unchanging shape, oscillating to and fro with amplitude b. It is known as a *coherent state* and is of special interest as it is the form of Ψ which results from applying a uniform time-varying force to a vibrator initially in its ground state. We 21 shall show this in due course; meanwhile let us note some of its characteristics by resolving it into its stationary components at time $\tau = 0$, when $\theta = 0$ and the spatial part of Ψ is $\pi^{-\frac{1}{4}} \mathrm{e}^{-\frac{1}{2}(\xi - b)^2}$. Then, if

$$\pi^{-\frac{1}{4}} \mathrm{e}^{-\frac{1}{2}(\xi - b)^2} = \sum_n a_n \psi_n,$$

a_n is determined by multiplying by ψ_n and integrating:

$$a_n = \int_{-\infty}^{\infty} \pi^{-\frac{1}{4}} \mathrm{e}^{-\frac{1}{2}(\xi - b)^2} \psi_n \,\mathrm{d}\xi = (2^n \pi n!)^{-\frac{1}{2}} \mathrm{e}^{-\frac{1}{4}b^2} \int_{-\infty}^{\infty} \mathrm{e}^{-\xi^2} H_n(\xi)\, \mathrm{e}^{\xi^2 - (\xi - \frac{1}{2}b)^2} \,\mathrm{d}\xi, \tag{19}$$

from (3) and (4). The rearrangement of the exponent shown in the last integral allows the use of a valuable identity concerning Hermite polynomials:

$$\mathrm{e}^{-\xi^2 - (\xi - \eta)^2} = \sum_l H_l(\xi)\eta^l / l! \tag{20}$$

The integral in (19) may therefore be written:

$$\begin{aligned}\int_{-\infty}^{\infty} \mathrm{e}^{-\xi^2} H_n\, \mathrm{e}^{\xi^2 - (\xi - \frac{1}{2}b)^2} \,\mathrm{d}\xi &= \sum_l (\tfrac{1}{2}b)^l / l! \int_{-\infty}^{\infty} \mathrm{e}^{-\xi^2} H_n H_l \,\mathrm{d}\xi \\ &= \sum_l (\tfrac{1}{2}b)^l / l! \int_{-\infty}^{\infty} (\psi_n \psi_l / \mathcal{N}_n \mathcal{N}_l) \,\mathrm{d}\xi \\ &= (\tfrac{1}{2}b)^n / \mathcal{N}_n^2 n!,\end{aligned}$$

on account of the orthogonality of the ψ_n. Hence

$$a_n = b^n \,\mathrm{e}^{-\frac{1}{4}b^2} / (2^n n!)^{\frac{1}{2}}, \tag{21}$$

and, from (14), the full form of $\Psi(\xi, \tau)$ follows:

$$\Psi(\xi, \tau) = \mathrm{e}^{-\frac{1}{4}b^2-\frac{1}{2}i\tau} \sum_n [b^n/(2^n n!)^{\frac{1}{2}}]\psi_n\, \mathrm{e}^{-in\tau}. \tag{22}$$

It is readily verified that (15) holds, for

$$\sum a_n a_n^* = \mathrm{e}^{-\frac{1}{2}b^2} \sum_n (\tfrac{1}{2}b^2)^n/n!,$$

and the summation is just the series expansion of $\mathrm{e}^{\frac{1}{2}b^2}$; hence $\Sigma\, a_n a_n^* = 1$.

A classical oscillator whose ξ-amplitude is b has energy $\frac{1}{2}b^2(\hbar\omega_0)$, and if we write $\frac{1}{2}b^2$ as n_c, n_c is to be interpreted as the number of quanta needed to supply the energy of the corresponding classical oscillator. The probability of finding the quantized oscillator in its nth state, with n quanta in addition to the zero-point energy, is $a_n a_n^*$ or $n_c^n\, \mathrm{e}^{-n_c}/n!$, and the mean number of quanta then follows:

$$\langle n \rangle = \mathrm{e}^{-n_c} \sum_n n n_c^n/n! = n_c.$$

When n_c is large a good approximation to $a_n a_n^*$ is obtained by use of Stirling's formula, $\ln n! \sim n \ln n - n$. Then if x is written for $n - n_c$,

$$\ln(a_n a_n^*) \sim -(n_c + x)\ln(1 + x/n_c) + x,$$

and the desired approximation follows by expanding $\ln(1 + x/n_c)$ as $x/n_c - x^2/2n_c^2$. To this order

$$a_n a_n^* \propto \mathrm{e}^{-(n-n_c)^2/2n_c}. \tag{23}$$

The energy levels that are significantly occupied lie in a range of about $n_c^{\frac{1}{2}}$, much smaller than n_c when the oscillator is highly excited. The energy of the oscillator in this non-stationary state is a constant of the motion and takes the same value as for a classical particle moving with the centroid of the wave-packet, apart from the half-quantum of zero-point energy which may be thought of as that required to confine the particle in the wave-packet; in fact the function $\pi^{-\frac{1}{4}}\, \mathrm{e}^{-\frac{1}{2}\xi^2}$ is just the ground state wave-function as given in (5). Other close parallels with the classical solution may also be noted – the energy of excitation (i.e. $E - \frac{1}{2}\hbar\omega_0$) is periodically exchanged between potential and kinetic forms, and the mean displacement $\langle \xi \rangle$ oscillates with simple harmonic motion. These two results are not peculiar to the particular solution chosen here, but are quite general as we now demonstrate. They arise from the specially simple form of the energy level structure (evenly spaced) and of the wave-functions, and are matched in virtually no other physical system.

Potential and kinetic energy

To show the exchange of energy consider the wave-function at time τ_0 expressed as a Fourier sum:

$$\Psi(\xi, \tau_0) = \sum_n a_n \psi_n.$$

Then one-quarter of a cycle later, at $\tau_0+\frac{1}{2}\pi$, each component will be multiplied by a phase-factor $e^{-\frac{1}{2}in\pi}$, so that

$$\Psi(\xi, \tau_0+\tfrac{1}{2}\pi)=\sum_n(-i)^n a_n\psi_n.$$

The potential energy in reduced units is $\langle\xi^2\rangle$ and hence

$$V(\tau_0)=\sum_m\sum_n\int_{-\infty}^{\infty} a_m^*\psi_m\xi^2 a_n\psi_n\,d\xi.$$

Similarly $$V(\tau_0+\tfrac{1}{2}\pi)=\sum_m\sum_n\int_{-\infty}^{\infty}(-i)^{n-m}a_m^*\psi_m\xi^2 a_n\psi_n\,d\xi.$$

Now of all the terms in these double summations, the alternating even and odd symmetries of the ψ_n ensure that the only terms that matter are those for which $n-m$ is even, but in fact out of all the integrals only those for which $n-m=0$ or ±2 are non-vanishing:

$$M^{(2)}_{n,n-2}\equiv\int_{-\infty}^{\infty}\psi_n\xi^2\psi_{n-2}\,d\xi=\tfrac{1}{2}[n(n-1)]^{1/2}$$

and $$M^{(2)}_{n,n}\equiv\int_{-\infty}^{\infty}\psi_n\xi^2\psi_n\,d\xi=n+\tfrac{1}{2}. \qquad (24)$$

Hence $$V(\tau_0)=\sum_n\{a_na_n^*(n+\tfrac{1}{2})+2\,\mathrm{Re}\,[a_n^*a_{n-2}]M^{(2)}_{n,n-2}\},$$

while $$V(\tau_0+\tfrac{1}{2}\pi)=\sum_n\{a_na_n^*(n+\tfrac{1}{2})-2\,\mathrm{Re}\,[a_n^*a_{n-2}]M^{(2)}_{n,n-2}\}.$$

Now the total energy ε is $\Sigma_n a_na_n^*(2n+1)$, just twice the first term, from which it immediately follows that

$$V(\tau_0+\tfrac{1}{2}\pi)=\varepsilon-V(\tau_0)=T(\tau_0), \qquad (25)$$

where $T(\tau_0)$ is the kinetic energy at τ_0. Every quarter-cycle the partitioning of the energy between potential and kinetic is interchanged. It may also be noted that the alternation of odd and even ψ_n ensures that $\Psi(\xi)\rightleftharpoons\Psi(-\xi)$ every half-cycle.

This is equally true for a classical harmonic vibrator, but there is one significant difference. It is possible to choose τ_0 for the classical vibrator so that the energy is entirely potential or entirely kinetic; there is complete interchange during each cycle. With the quantized vibrator, on the other hand, there are no states in which either potential or kinetic energy is entirely absent. Thus the function (16) at time $\tau=0$ takes the form $\pi^{-\frac{1}{4}}e^{-\frac{1}{2}(\xi-b)^2}$ in which the displacement is greatest; V has its maximal value of b^2 and T its minimal value of $\frac{1}{2}$, the zero-point kinetic energy associated with the constant-phase Gaussian wave-packet. One quarter-cycle later, Ψ is an undisplaced wave-packet and V is minimal, at $\frac{1}{2}$, while the phase-variation across the wave-packet contributed by θ now confers on T its maximal value of b^2. There is complete interchange between T and V except for the zero-point energy, $\frac{1}{2}$. By contrast, in a stationary state $T=V$ at all times, and no interchange takes place. The restricted interchange permitted by (25) may be modelled in a classical system by an ensemble of identical vibrators rather than a single one. If the vibrators have different phases each exhibits complete interchange, but there is never a moment when all are purely potential or purely kinetic; and if the phases are evenly distributed over all possible values the total potential and kinetic energies remain unchanged with time, just as in the stationary state of a quantized vibrator. As we shall see in a later section, this is by no means the only way an ensemble of classical vibrators can be made to model the quantized vibrator.

21

It is characteristic of the total potential and kinetic energies of the classical ensemble not simply that they obey (25) but that each oscillates sinusoidally at twice the natural frequency:

e.g. $$V = \bar{V} + \Delta V \cos 2\tau, \qquad T = T_0 - \Delta V \cos 2\tau.$$

The result quoted in (24) leads to the same result for the quantized vibrator. For

$$\Psi(\xi, \tau_0 + \tau) = \sum_n a_n \psi_n \, \mathrm{e}^{-\mathrm{i}(n+\frac{1}{2})\tau},$$

and therefore $$V(\tau_0+\tau) = \sum_m \sum_n \int_{-\infty}^{\infty} a_m^* \psi_m \xi^2 a_n \psi_n \, \mathrm{e}^{-\mathrm{i}(n-m)\tau} \, \mathrm{d}\xi.$$

On multiplying out and performing the integrations the only surviving terms are constants, when $m = n$, or have time-dependence as $\mathrm{e}^{\pm 2\mathrm{i}\tau}$ when $n - m = \pm 2$. It is, of course, the regular spacing of the levels that ensures that all values of n generate the same fundamental frequency, 2 in this case; and it is the vanishing of $M_{n,m}^{(2)}$ for any $|n-m|$ greater than 2 that eliminates all harmonics.

Classical behaviour of $\langle \xi \rangle$

A similar result holds for $\langle \xi \rangle$ as for $\langle \xi^2 \rangle$:

$$\langle \xi \rangle = \sum_m \sum_n \int_{-\infty}^{\infty} a_m^* \psi_m \xi a_n \psi_n \, \mathrm{e}^{-\mathrm{i}(n-m)\tau} \, \mathrm{d}\xi.$$

Now symmetry permits only odd values of $|n-m|$ to contribute, and in fact only when $|n-m| = 1$ does the integral not vanish:

$$M_{n-1,n}^{(1)} = M_{n,n-1}^{(1)} \equiv \int_{-\infty}^{\infty} \psi_n \xi \psi_{n-1} \, \mathrm{d}\xi = (n/2)^{\frac{1}{2}}. \qquad (26)$$

Consequently $\langle \xi \rangle$ contains terms only in $\mathrm{e}^{\pm \mathrm{i}\tau}$ and oscillates sinusoidally, without a constant term, at the natural frequency.

This last result may be proved, without Fourier analysis, directly from Schrödinger's equation by use of Ehrenfest's theorem[5], which is valid for any one-dimensional system:

$$m\frac{\mathrm{d}}{\mathrm{d}t}\langle x\rangle=\langle p\rangle \quad \text{and} \quad \frac{\mathrm{d}}{\mathrm{d}t}\langle p\rangle=-\langle \text{grad } V\rangle,$$

so that

$$m\ \mathrm{d}^2\langle x\rangle/\mathrm{d}t^2=-\langle \text{grad } V\rangle. \tag{27}$$

It is not necessary that the potential V should be constant – it may also include the potential of some time-varying force. The close similarity between (27) and the classical expression of Newton's laws of motion is rather deceptive, in that the mean position $\langle x\rangle$ responds not to $F(\langle x\rangle)$, the force at this one point, but to $\langle F(x)\rangle$, the weighted average value of the force over all possible positions of the particle. Thus (27) describes perfectly well the tunnelling of a particle through a barrier, an impossible process in classical mechanics. If, however, grad V is either constant or contains in its series expansion no term of higher than first order, $\langle x\rangle$ does indeed follow the classical law. For when $-\text{grad } V=f_0(t)+xf_1(t)$, (27) takes the form

$$m\ \mathrm{d}^2\langle x\rangle/\mathrm{d}t^2=f_0(t)+\langle x\rangle f_1(t)=F(\langle x\rangle), \tag{28}$$

so that $\langle x\rangle$ behaves exactly like the displacement of a classical particle. In particular, if $f_0=0$ and f_1 is a negative constant, $\langle x\rangle$ executes harmonic motion in the Hooke's law potential, as proved above. And if now a uniform force, represented by non-vanishing $f_0(t)$, is applied $\langle x\rangle$ develops in exactly the same way as a classical oscillating particle subjected to $f_0(t)$. It may also be remarked that if $f_0=0$ and f_1 includes a time-varying component, (28) describes an oscillator parametrically excited by variations of the force constant, and $\langle x\rangle$ in this case also develops along classical lines. This strict parallelism between quantal and classical does not extend to variables other than $\langle x\rangle$, but the idea can be carried somewhat further by introducing an ensemble of classical oscillators which collectively represent the probability distribution of x and its response to a uniform force or to parametric excitation. For this development we shall need to know how a quantized oscillator responds to a sharp impulsive force, and as this is a matter of wider significance than the present application we shall establish the required result in general terms, though leaving the wider discussion until a later chapter.

86

Impulse-response of a quantized system

A particle moving in a constant three-dimensional potential $V(\boldsymbol{r})$ and subjected to a varying force $F(t)$ in the x-direction, independent of $\boldsymbol{r}$, obeys the Schrödinger equation

$$(\hbar^2/2m)\nabla^2\Psi-(V-xF)\Psi=-\mathrm{i}\hbar\partial\Psi/\partial t. \tag{29}$$

Now let F be a sharp impulse $P\delta(t-t_0)$ applied at time t_0. During the infinitesimal interval of its application the term in F completely dominates the left-hand side, so that the immediate effect of the impulse is obtained by integrating the equation:

$$\partial\Psi/\partial t = \mathrm{i}xF\Psi/\hbar.$$

If Ψ_- and Ψ_+ are the wave-functions immediately before and after the impulse,

$$\Psi_+ = \Psi_- \,\mathrm{e}^{\mathrm{i}x\int F\,\mathrm{d}t/\hbar} = \Psi_- \,\mathrm{e}^{\mathrm{i}xP/\hbar}. \tag{30}$$

The magnitude of Ψ is unchanged, but a phase shift proportional to x is imposed. This result holds for any strength of impulse, but when the impulse is weak it may be convenient to use the approximate form:

$$\Delta\Psi = \Psi_+ - \Psi_- \sim \mathrm{i}xP\Psi/\hbar \tag{31}$$

It is worth noting that (31) follows by direct integration of (29) over a short interval Δt, in which F is taken as constant:

$$\Delta\Psi = (\mathrm{i}x/\hbar)F\Psi\Delta t + \frac{\mathrm{i}}{\hbar}\left(\frac{\hbar^2}{2m}\nabla^2\Psi - V\Psi\right)\Delta t.$$

The second term represents the development of Ψ in the absence of a disturbing force F, while the first term is the change due to the impulse $F\Delta t$. The linearity of the differential equation for Ψ ensures that the two processes take place independently.†

To determine the impulse response function let us start with the system in a non-stationary state represented by $\Sigma_l\, a_l\psi_l$ at time $t=0$. If a weak impulse P is applied at t_0, when $\Psi(t_0) = \Sigma_l\, a_l\psi_l\,\mathrm{e}^{-\mathrm{i}E_lt_0/\hbar}$, the resulting change in the wave-function follows from (31): I.106

$$\Delta\Psi(t_0) = (\mathrm{i}xP/\hbar)\sum_l a_l\psi_l\,\mathrm{e}^{-\mathrm{i}E_lt_0/\hbar}. \tag{32}$$

This change in Ψ can be represented as a change Δa_l in each a_l, and if the system remains undisturbed thereafter the new set of a_l stays constant. We therefore evaluate the coefficients at t_0, writing

$$\Delta a_m = \int_{-\infty}^{\infty} \psi_m^*\,\mathrm{e}^{\mathrm{i}E_mt_0/\hbar}\Delta\Psi(t_0)\,\mathrm{d}\boldsymbol{r} = (\mathrm{i}P/\hbar)\sum_l a_lM_{m,l}^{(1)}\,\mathrm{e}^{\mathrm{i}(E_m-E_l)t_0/\hbar}, \tag{33}$$

in which $M_{m,l}^{(1)}(=M_{l,m}^{(1)*})$ is written for the matrix element $\int_{-\infty}^{\infty}\psi_m^*x\psi_l\,\mathrm{d}\boldsymbol{r}$.

† The linearity in Ψ of Schrödinger's equation, for any $V(\boldsymbol{r})$, implies that if any two functions, Ψ_1, and Ψ_2, are solutions then so is $a_1\Psi_1 + a_2\Psi_2$. This should not be confused with the linearity in x exhibited by the equation of motion of a classical harmonic vibrator, $m\ddot{x} + \mu x = F(t)$. The latter has the property that the response to two forces, $F_1(t)$ and $F_2(t)$, applied simultaneously, is the sum of the responses to each separately. It is not true that the development of Ψ can be similarly dissected into the responses to different components of the applied force, even for the harmonic vibrator (though in this case $\langle x\rangle$ has this property, as (28) shows).

Hence after a further interval t,

$$\Psi_+(t_0+t)=\sum_m (a_m+\Delta a_m)\psi_m\,\mathrm{e}^{-\mathrm{i}E_m(t_0+t)/\hbar}$$

$$=\Psi_-(t_0+t)+(\mathrm{i}P/\hbar)\sum_m\sum_l a_l M^{(1)}_{m,l}\psi_m\,e^{-\mathrm{i}(E_l t_0+E_m t)/\hbar}, \qquad (34)$$

where $\Psi_-(t_0+t)=\Sigma_l\, a_l\psi_l\,\mathrm{e}^{-\mathrm{i}E_l(t_0+t)/\hbar}$, representing the development of Ψ which would have taken place without the impulse.

We may now calculate the variation of any observable, and shall concentrate on $\langle x\rangle$, evaluated as $\int_{-\infty}^{\infty}\Psi_+^* x\Psi_+\,\mathrm{d}x$. From (34) it is seen that $\langle x\rangle$ contains a term in Ψ_- independent of P, which is the variation in the absence of an impulse, a term proportional to P which is the required impulse-response of $\langle x\rangle$, and a term in P^2 which we ignore since (31) is valid only to first order. Representing the middle term as $\langle\Delta x\rangle$, we have for the impulse-response function

$$h(t)\equiv\langle\Delta x\rangle/P=(i/\hbar)\sum_l\sum_m\sum_n a_n^* a_l M^{(1)}_{n,m}M^{(1)}_{m,l}\,\mathrm{e}^{-\mathrm{i}[(E_l-E_n)t_0+(E_m-E_n)t]/\hbar}+\text{c.c.} \qquad (35)$$

Clearly there may be many frequencies present in the oscillations of $h(t)$, a typical example being $(E_m-E_n)/\hbar$ arising from the beating of the wavefunctions of the mth and nth states. Such a term may well have been present before the impulse, but of course $h(t)$ contains only the changes in its amplitude resulting from the impulse. If the lth state was originally present, its perturbation by P results in a change Δa_m in the amplitude of the mth state, to an extent determined by $a_l M^{(1)}_{m,l}$; the beating of the mth and the nth states, which yields an amplitude in $h(t)$ proportional to $a_m a_n M^{(1)}_{m,n}$, is to first order changed by $\Delta a_m\cdot a_n M^{(1)}_{m,n}$, so that the change in the amplitude at frequency $(E_m-E_n)/\hbar$ involves three levels, as indicated by the triple summation in (35).

Let us now take a special case, where the system is initially in the nth stationary state, so that $a_n=1$ and all other a_l vanish. Then the triple summation collapses to a single summation, t_0 may be taken as zero (since for a stationary state all times are equivalent), and

$$h(t)=(\mathrm{i}/\hbar)\sum_m |M^{(1)}_{m,n}|^2\,\mathrm{e}^{-\mathrm{i}(E_m-E_n)t/\hbar}+\text{c.c.} \qquad (36)$$

Writing ω_{mn} for $(E_m-E_n)\hbar$ and $f_{m,n}$ for $(2m\omega_{mn}/\hbar)|M^{(1)}_{m,n}|^2$, we have†

I.107

$$h(t)=\sum_m (f_{mn}/m\omega_{mn})\sin\omega_{mn}t. \qquad (37)$$

In this form the response of the system bears a strong resemblance to that of a set of classical oscillators of different frequencies, ω_{mn}, and masses

† Note that m is used both for the mass of the particle and as an index of the stationary state. In the latter sense it is not used except as a subscript.

m/f_{mn}; and this in spite of the real system having no resemblance to an oscillator – its character has indeed not been specified. This result will form the basis of a later general discussion of the transitions between states induced by a time-varying force. For the present we confine our attention to applying it to a harmonic vibrator.

It is well to note, however, one general point arising from (35) as a warning against using the impulse-response function too casually. The presence of t_0 in the expression for $h(t)$ shows that, in contrast to a classical vibrator, the quantized system will respond differently at different times if it is not in a stationary state. This is clearly because each Δa_m has phase as well as amplitude, so that when it beats with a_n the outcome will depend on the relative phase of a_m and the particular a_l whose perturbation generated Δa_m; hence the presence of $(E_l - E_n)t_0/\hbar$ in the phase. The only occasions when a time-independent response function may be used safely is when the system is never driven significantly far from a pure stationary state; or when it is in a mixture of stationary states but with unknown phase relationships, and all that is required is the average response taken over all possible phases. Phase-averaging of (35) eliminates all terms for which $m \neq l$, yielding a slightly generalized form of (37):

$$h(t) = \sum_m \sum_n a_n^* a_n (f_{mn}/m\omega_{mn}) \sin \omega_{mn} t. \tag{38}$$

Each occupied level in the initial state responds independently in proportion to its occupation probability $a_n^* a_n$, and the response to a time-varying force may be expressed as an integral over impulse-responses of this standard form provided the $a_n^* a_n$ are not greatly changed. This is the condition for treating the response as linear. It happens, as (28) shows, that even for strong perturbations the harmonic vibrator behaves linearly so far as $h(t)$, the response of $\langle x \rangle$, is concerned; but this is not obvious from the present argument.

The form of $h(t)$ for a harmonic vibrator initially in the nth state is especially simple, since according to (26) $M_{m,n}^{(1)}$ vanishes unless $m = n \pm 1$. In laboratory variables $|M_{n-1,n}^{(1)}|^2 = (\hbar/2m\omega_0)n$ and $|M_{n+1,n}^{(1)}|^2 = (\hbar/2m\omega_0)(n+1)$, so that $f_{n-1,n} = -n$ and $f_{n+1,n} = n+1$. Hence **80**

$$h(t) = [(n+1)/m\omega_0] \sin \omega_0 t - [n/m\omega_0] \sin \omega_0 t = (\sin \omega_0 t)/m\omega_0, \tag{39}$$

which is, as expected, the same as the classical expression for the impulse-response function. It is the absence from (39) of the initial value of n that makes possible (though it does not prove) the application of this result to any strength of perturbation. The oscillatory response (39) is undamped, since we have introduced no damping mechanism. The problem of damping will be raised in chapter 16 and subsequently.

To proceed from the impulse-response to the response to a time-varying force, we refer to (33) to write down how the amplitudes a_n develop, treating the force $F(t)$ as a sequence of impulses:

$$\dot{a}_n = (\mathrm{i}F(t)/\hbar)\sum_l a_l M^{(1)}_{n,l}\,\mathrm{e}^{-\mathrm{i}\omega_{nl}t}. \quad (40)$$

This is a general result, which we now apply to the harmonic vibrator, returning at this point to reduced variables, with Schrödinger's time-dependent equation in the form

$$\partial^2\Psi/\partial\xi^2 - (\xi^2 - 2\xi\phi)\Psi = -2\mathrm{i}\partial\Psi/\partial\tau,$$

in which $\phi(\tau) = F(t)/(m\hbar\omega_0^3)^{1/2}$. Then (40) takes the form

$$\mathrm{d}a_n/\mathrm{d}\tau = (\mathrm{i}\phi/\sqrt{2})[n^{\frac{1}{2}}a_{n-1}\,\mathrm{e}^{\mathrm{i}\tau} + (n+1)^{\frac{1}{2}}a_{n+1}\,\mathrm{e}^{-\mathrm{i}\tau}]. \quad (41)$$

In particular, if ϕ is a sinusoidally varying force, $\mathrm{Re}\,[\phi\,\mathrm{e}^{-\mathrm{i}\omega_0(1+\delta)t}]$, we write it as $\frac{1}{2}(\phi^*\,\mathrm{e}^{\mathrm{i}(1+\delta)\tau} + \phi\,\mathrm{e}^{-\mathrm{i}(1+\delta)\tau})$ to give four terms in (41):

$$\mathrm{d}a_n/\mathrm{d}\tau = (i/2\sqrt{2})[n^{\frac{1}{2}}a_{n-1}(\phi^*\,\mathrm{e}^{\mathrm{i}(2+\delta)\tau} + \phi\,\mathrm{e}^{-\mathrm{i}\delta\tau}) + (n+1)^{\frac{1}{2}}a_{n+1}(\phi^*\,\mathrm{e}^{\mathrm{i}\delta\tau} + \phi\,\mathrm{e}^{-\mathrm{i}(2+\delta)\tau})].$$

When $\delta \ll 1$, close to resonance, the rapidly alternating terms at frequency $2+\tau$ provide only a faint ripple on a_n and may be ignored, so that

$$\mathrm{d}a_n/\mathrm{d}\tau = (\mathrm{i}/2\sqrt{2})[\phi n^{\frac{1}{2}}a_{n-1}\,\mathrm{e}^{-\mathrm{i}\delta\tau} + \phi^*(n+1)^{\frac{1}{2}}a_{n+1}\,\mathrm{e}^{\mathrm{i}\delta\tau}]. \quad (42)$$

As δ goes to zero the variations in a_n become steadily larger; the wave-packet, following the classical motion, oscillates with an amplitude b that beats at frequency δ, the peak amplitude varying inversely as δ. It does, in fact, take the form (16) with b now a function of time. Correspondingly, as (23) shows, the values of n which are most sharply excited oscillate at frequency δ; a plot of $a_n^* a_n$ against n at various times would show the system starting, say, in its lowest state, climbing up the ladder of levels until a rather wide range $(\sim n_c \pm n_c^{\frac{1}{2}})$ was sparsely occupied, and then collapsing back again to the ground state, to repeat the pattern unchanged and indefinitely so long as the same sinusoidal force was applied. Only at resonance, when $\mathrm{e}^{\pm\mathrm{i}\delta\tau} = 1$, does the climb proceed unabated.

Critical phenomena

It has been possible to describe the solution of the set of equations represented by (42) in these terms without a detailed calculation because the full solution (21) is already known. We therefore know, among other things, that so long as $\delta \neq 0$ the system does not diverge – there is always some value of n above which virtually no excitation is found at any time. In general, however, there are more than two terms in (42), although most will be far from resonance with any single excitation frequency. It is then hard to be sure that the response will be convergent, or that the pattern of development of the a_n will vary smoothly with the applied frequency. There are, indeed, simple examples exhibiting critical phenomena, which would be difficult to discern from the structure of the equations corresponding to (42). Such a case occurs with parametric excitation of a harmonic vibrator. **I.285**

As already noted, following (28), under the influence of parametric excitation by periodic variation of the force constant, $\langle x \rangle$ develops in the classical way. This means that for a given strength of excitation there is a band of frequencies around $2\omega_0$ in which an initial small vibration is amplified, growing exponentially without limit if the system is truly harmonic. Thus as the frequency is moved towards $2\omega_0$ the variations of a_n remain within bounds, but at a certain critical frequency they become divergent. We may ask whether the equations corresponding to (42) in this case exhibit any feature that allows the critical phenomenon to be discerned, and the actual critical frequency to be calculated. Unhappily the answer seems to be no, but let us derive the equations.

If the force constant is modulated at a frequency of $2r\omega_0$, r being close to unity, (13) takes the form:

$$\partial^2\Psi/\partial\xi^2 - \xi^2(1+\beta\cos 2r\tau)\Psi = -2\mathrm{i}\partial\Psi/\partial\tau, \tag{43}$$

β being a measure of the strength of the parametric excitation. We seek a solution in the form (14) (with l replacing n) where now the a_l may vary with time. Substitute (14) in (43), multiply by $\psi_n\,\mathrm{e}^{\mathrm{i}n\tau}$ and integrate over ξ; then

$$\mathrm{d}a_n/\mathrm{d}\tau = -\tfrac{1}{2}\mathrm{i}\beta\cos r\tau\sum_l a_l M^{(2)}_{l,n}\,\mathrm{e}^{\mathrm{i}(n-l)\tau},$$

the matrix elements $M^{(2)}_{l,n}$ being given in (24); only those for which $l-n=0$ or ± 2 are non-vanishing. When their values are substituted and the rapidly varying terms eliminated, as was done with (42), the remainder take the form:

$$\mathrm{d}a_n/\mathrm{d}\tau = -\tfrac{1}{8}\mathrm{i}\beta[(n+2)^{\frac{1}{2}}(n+1)^{\frac{1}{2}}a_{n+2}\,\mathrm{e}^{2\mathrm{i}\delta\tau} + n^{\frac{1}{2}}(n-1)^{\frac{1}{2}}a_{n-2}\,\mathrm{e}^{-2\mathrm{i}\delta\tau}], \tag{44}$$

in which $\delta = r-1$. This differs from (42) in that the coefficients in the square brackets vary as n, rather than $n^{\frac{1}{2}}$, at large values of n. Thus high on the ladder of levels there is greater mobility and one need not be surprised to learn that when δ is small enough (44) may represent an uncontrolled escalation, while (42) only shows this behaviour when $\delta = 0$.

It only requires a small modification in (42) to create a comparable situation. A slightly anharmonic vibrator, if nearly lossless, shows a discontinuous response as the driving force is tuned through resonance. In one direction the amplitude climbs to a high value and suddenly plunges, while in the reverse direction it remains low until with equal suddenness it soars. Presumably the form of (42) for this case is almost unchanged, except that the levels are not quite evenly spaced, and δ must be regarded as varying with n. The discontinuous switches are associated with δ going through zero at some non-zero value of n, and the system either climbing or descending catastrophically through this region of the ladder of energy **I.250** levels. However, as with parametric excitation, it is no easy matter to quantify the critical phenomenon when (41) or its equivalent is taken as the starting point. Examples of such critical behaviour may be only rarely encountered in quantum phenomena, but the very possibility of their undetected presence is worrying. This example perhaps demonstrates more clearly than any other the difference between the power of classical and quantum mechanics to cope with problems out of the ordinary. If for no other reason, classical treatments should be valued for the warning they give of potential dangers lurking in the mathematical jungle of quantum mechanics.

The equivalent classical ensemble (σ-representation)

We return to the undisturbed vibrator in a non-stationary state, with a wave-function that obeys (13). The particular solution (16) provides the simplest example of a general property of solutions of (13): if $\Psi_0(\xi, \tau)$ is any solution, then $\Psi(\xi, \tau)$ is also a solution, where

$$\Psi(\xi, \tau) = \Psi_0(\xi - b\cos\tau, \tau)\exp\left[-\mathrm{i}(b\xi\sin\tau_1 - \tfrac{1}{4}b^2\sin 2\tau_1 + c)\right]. \quad (45)$$

Here τ_1 is written for $\tau + \tau_0$; τ_0, b and c are arbitrary real constants. The truth of this proposition may be tested by substitution in (13). The probability distribution, $\mathscr{P}(\xi, \tau)$, defined as $\Psi\Psi^*$, contains nothing of the rather complicated phase variation of (45):

$$\mathscr{P}(\xi, \tau) = \mathscr{P}_0(\xi - b\cos\tau, \tau), \quad (46)$$

where $\mathscr{P}_0 = \Psi_0\Psi_0^*$. The probability distribution corresponding to Ψ develops periodically in the same way as that corresponding to Ψ_0, but also vibrates bodily with amplitude b. We now show that the effect of a time-varying uniform force on a system initially described by Ψ_0 is to change Ψ_0 into Ψ, with the form (45), by progressive changes in the parameters b, c and τ_0. It is enough for this purpose to show that a system described by (45), when acted upon by an arbitrary impulse, merely suffers certain changes in these parameters. It will then follow that after the impulse, and therefore through the sequence of changes consequent upon applying a varying force, $\mathscr{P}(\xi, \tau)$ continues to go through exactly the same periodic variation while vibrating to and fro with a continuously changing amplitude, b.

To demonstrate this result, consider (45) as representing Ψ_- just before the impulse P; then immediately after, according to the exact result (30),

$$\Psi_+(\xi, \tau) = \Psi_0(\xi - b\cos\tau, \tau)\exp\left[-\mathrm{i}(b\xi\sin\tau_1 - \tfrac{1}{4}b^2\sin 2\tau_1 + c - p\xi)\right],$$

in which p is the reduced form of $P(Px/\hbar = p\xi)$. This expression for Ψ_+ may now be rewritten in the same form as (45) by changing the constants to b', c' and τ_0', satisfying the relations:

$$b \cos \tau_1 = b' \cos \tau_1', \qquad (47)$$

$$b \sin \tau_1 - p = b' \sin \tau_1', \qquad (48)$$

and

$$c - \tfrac{1}{4}b^2 \sin 2\tau_1 = c' - \tfrac{1}{4}b'^2 \sin 2\tau_1'. \qquad (49)$$

The change in c given by (49) does not concern us, though it may be remarked that it is represented in fig. 2(a) by the area of the shaded triangle. In the same diagram are shown the changes in b and τ_1, exactly the same as if they defined the clockwise-rotating vector describing a classical oscillator; the real part, the displacement, is unchanged by the impulse, but the imaginary part, the momentum, is increased by p.

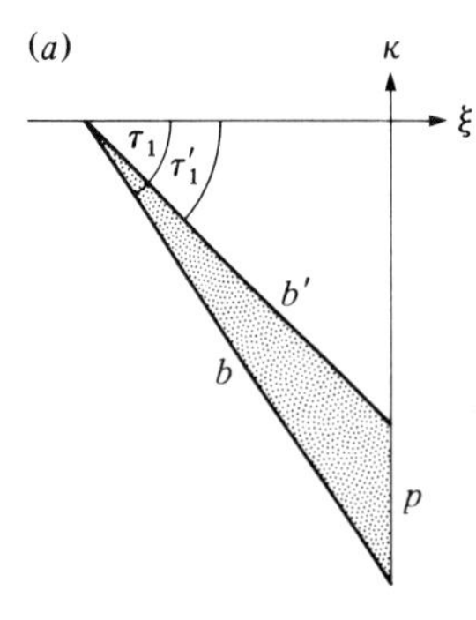

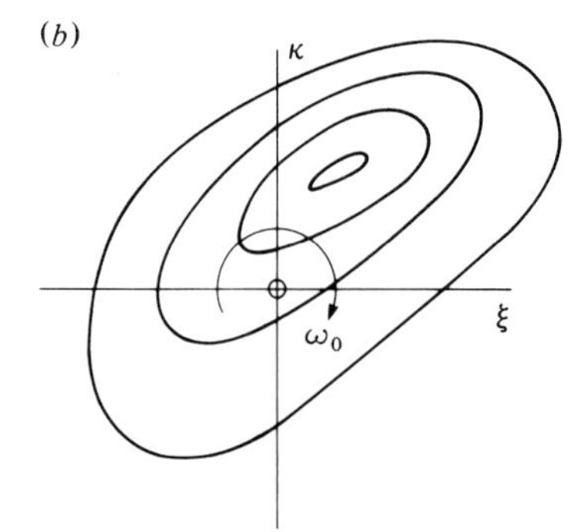

Fig. 2. (a) Illustrating the solution of (47)–(49). (b) Schematic diagram of the σ-representation, showing hypothetical contours of σ on the (ξ, κ)-plane.

We are now in a position to construct the classical model which will reproduce the behaviour of $\mathscr{P}(\xi, \tau)$ under the influence of a varying uniform force. Let us for the moment assume, as will be proved directly, that when no force is applied the periodic variations of $\mathscr{P}_0$ may always be represented by an ensemble of classical vibrators, defined by a two-dimensional density distribution $\sigma(\boldsymbol{R})$, where $\boldsymbol{R} = (\xi, \kappa)$, such as is shown schematically in fig. 2(b). The contours of σ are to be imagined drawn on a rigid lamina which spins at the oscillation frequency about the origin, so that at time τ the density is $\sigma(\xi \cos \tau - \kappa \sin \tau, \xi \sin \tau + \kappa \cos \tau)$. At any instant $\mathscr{P}_0$ is the projection of σ on to the ξ-axis:

$$\mathscr{P}_0(\xi, \tau) = \int_{-\infty}^{\infty} \sigma(\xi \cos \tau - \kappa \sin \tau, \xi \sin \tau + \kappa \cos \tau)\, \mathrm{d}\kappa. \qquad (50)$$

What the argument of the last paragraph showed was that the effect of an impulse may be represented by suddenly displacing the pattern of σ (i.e. the lamina) vertically through p, and then allowing the rotation to continue about the origin. This leaves the development of $\mathscr{P}$ unchanged, but superposes a bodily oscillation as described by a change of b and τ_1 in (46). We may think of σ as defining an ensemble of classical vibrators, each represented by a rotating vector, the heads of vectors having a density distribution σ. Since the ξ-component of each vector defines the instantaneous displacement of the corresponding oscillator, $\mathscr{P}_0(\xi, \tau)$ as defined by (50) is just the distribution of displacements at τ. And it may be remarked incidentally that the κ-components of the ensemble equally well describe the momentum-distribution in Ψ_0, and its variation with time; this explains the choice of κ to denote this coordinate. Finally, to complete the story, fig. 2(a) shows that if each member of the classical ensemble is subjected to the same impulse p, it responds in the required way to produce the modified form of σ; and what holds for an arbitrary impulse automatically holds for a time-varying force. Hence the σ-representation, which is the term we use from now on to refer to the classical ensemble set up to describe the initial probability distribution $\mathscr{P}_0$, and its cyclical evolution, will continue to give a correct description when the quantized system and its ensemble analogue are disturbed by an arbitrary uniform force.

Nothing has been said about the conditions that $\sigma(\boldsymbol{R})$ must satisfy in order to describe a possible distribution $\mathscr{P}(\xi, \tau)$. Obviously not every σ has this property (e.g. a strongly localized peak cannot describe the distribution arising from any real wave-function), but the restrictions on σ are hard to derive. For our purpose this is irrelevant, but it is not irrelevant that every possible $\mathscr{P}(\xi, \tau)$ can be represented by the appropriate σ. The proof is easy. Let σ at some instant be expressed as a two-dimensional Fourier integral,

$$\sigma = \iint A(k_\xi, k_\kappa)\, \mathrm{e}^{\mathrm{i}(k_\xi \xi + k_\kappa \kappa)}\, \mathrm{d}k_\xi\, \mathrm{d}k_\kappa.$$

Then $\mathscr{P}(\xi)$ at the same instant is $\int \sigma\, \mathrm{d}\kappa$ to which only Fourier components having $k_\kappa = 0$ contribute:

$$\mathscr{P}(\xi) \propto \int A(k_\xi, 0)\, \mathrm{e}^{\mathrm{i}k_\xi \xi}\, \mathrm{d}k_\xi.$$

Hence by the Fourier transform theory the instantaneous value of $\mathscr{P}(\xi)$ may be used to define A along the ξ-axis, and at any other time, when σ has turned to a new orientation, the variation of A along another radius is again uniquely determined by $\mathscr{P}$. It is therefore always possible to represent any $\mathscr{P}(\xi, \tau)$ by one, and one only, spinning σ. The fact that $\mathscr{P}$ reverses itself every half-cycle is an essential requirement.

For a stationary state σ has circular symmetry and its spinning produces no change. Thus for the unnormalized ground state in which $\Psi = \mathrm{e}^{-\frac{1}{2}(\xi^2 + \mathrm{i}\tau)}$, $\mathscr{P}(\xi)$ is $\mathrm{e}^{-\xi^2}$ and σ is e^{-R^2}, an isotropic two-dimensional Gaussian peak. If this narrow peak is displaced far from the origin and left to spin unchanged round a large circle it describes a coherent state of excitation. A stationary state of high energy, specified by a single n rather than the range expressed in (21), is by contrast represented by something like a narrow ring of constant amplitude all round. The mean level of $\mathscr{P}(\xi)$ follows closely the classical distribution, $1/\pi(\xi_0^2 - \xi^2)^{\frac{1}{2}}$, for an exact description of which σ must be wholly concentrated on a circle of radius ξ_0 and zero width. The lack of perfect sharpness at the edge, which we have already noted, means that the ring to which σ is confined must have a width of the order of $1/n^{\frac{1}{6}}$.

But this is not the whole story since $\mathscr{P}(\xi)$ oscillates, with n zeroes. To reproduce these oscillations the ring that generates the mean value must be supplemented by an extended pattern of relatively low-amplitude rings, on which σ alternates between positive and negative. We have no need to ask about the details of the pattern, but it is worth remarking that the presence of zeroes in $\mathscr{P}(\xi)$ makes negative values of σ inevitable. These have no classical interpretation; a negative density of vectors at ξ does not produce in $\mathscr{P}(\xi)$ the same contribution as a positive density at $-\xi$. The σ-representation is a convenient fiction, incapable of realization. We shall make considerable use of it later.

Energy imparted by an applied force

We shall now show that a uniform force with arbitrary time variation changes the energy of the harmonic vibrator in an essentially classical manner. It is sufficient to establish the result for an impulse acting on any state that can be represented on the $\sigma(\xi, \kappa)$ diagram. It follows from (11) that the potential energy $\langle V\rangle$ at any instant, measured in terms of $\frac{1}{2}h\omega_0$, is $\langle\xi^2\rangle$, while $\langle\kappa^2\rangle$ is the value $\langle V\rangle$ will acquire one-quarter of a cycle later when the pattern has spun through $\pi/2$. Hence, from (24), $\langle\kappa^2\rangle$ is the kinetic energy $\langle T\rangle$, and the total energy E is $\langle\xi^2+\kappa^2\rangle$, in accordance with the earlier statement associating κ in this representation with momentum. **9**
Writing $\xi^2+\kappa^2$ as R^2 we have that

$$\langle R^2\rangle = \bar{R}^2 + \langle(R-\bar{R})^2\rangle,$$

$\bar{R}$ being the co-ordinate of the centroid of the distribution. We have seen that an impulse shifts the distribution bodily in the κ-direction, so that the second term is unchanged. Thus we reach the required result, that the total energy changes by the same amount as if the impulse were applied to a classical oscillator at $\bar{R}$.

An alternative proof, of which the details need not be given, involves calculating $\langle T\rangle$ as $-(\hbar^2/2m)\int\psi^*(\mathrm{d}^2\psi/\mathrm{d}x^2)\,\mathrm{d}x$, and determining the change $\Delta\langle T\rangle$ when an impulse P replaces ψ by $\psi\,\mathrm{e}^{\mathrm{i}xP/\hbar}$, as in (30). Then $\Delta\langle T\rangle$ is found in general to be $P\langle\dot{x}\rangle+P^2/2m$, the same as the classical result in the special case of a harmonic vibrator for which we know $\langle x\rangle$ evolves classically under the influence of a uniform force.

At this point one may begin to worry that everything seems to conform too closely to the classical model, although we know that there are highly significant differences, which were responsible for the original discovery of the quantum. The mean energy of a vibrator in equilibrium with a gas at temperature T is not the classical equipartition energy $k_\mathrm{B}T$ but is given by Planck's expression $\hbar\omega_0/(\mathrm{e}^{\hbar\omega_0/k_\mathrm{B}T}-1)$, which is less than $k_\mathrm{B}T$ at all T.† If we suppose the gas atoms, colliding with the vibrator, to exert on it

† Planck's expression does not contain the zero-point energy $\frac{1}{2}\hbar\omega_0$, and is therefore the mean excitation energy, not the total energy. If the zero-point energy is included, the mean total energy is $\frac{1}{2}\hbar\omega_0\coth(\hbar\omega_0/2k_\mathrm{B}T)$, which is greater than $k_\mathrm{B}T$, its asymptotic value as $T\to\infty$.

a time-varying force we might take the argument just developed as good reason to expect the transfer of energy to be indifferent as to the classical or quantal character of the vibrator. The mistake lies in thinking that it reacts to the collision in the same way as to a specified impulse. This is not to imply that it is the non-uniformity of the force that is the sole cause of the difference; something more fundamental is involved. When two atomic systems interact, the consequences can only be determined by solving the Schrödinger equation for the two together; it is invalid to imagine that each acts on the other in some specified way, and that the **163** two-body problem can be thus dissected into two one-body problems. An obvious illustration is provided by a vibrator in its ground state being struck by an atom moving so slowly that its energy is less than $\hbar\omega_0$. There is no objection to some of this energy being transferred in a classical collision, nor to $\langle E \rangle$ increasing if the dissection of the quantum system is attempted, and the impact treated as a weak time-dependent force; but we know that in the correct solution there must be no chance whatever of finding the vibrator in an excited state afterwards. This fundamental change in the dynamics is enough to dispel any expectation that equipartition of energy should apply.

The objection on the grounds of energy conservation is less serious when the colliding body is massive, for it may be moving very slowly and yet carry energy much in excess of $\hbar\omega_0$. Presumably, though we have not proved this, when the energy transfer calculated classically is much less than the energy of the colliding body it may be assumed that the motion of the body is virtually unperturbed and that the vibrator will react as to **67** a prescribed force. We shall meet a similar problem in chapter 16, where we shall find it impermissible to pretend that when two quantized vibrators are coupled together each responds independently to the time-varying force exerted by the other. Yet if one of the vibrators is very small and at a low level of excitation, while the other is large and in a high coherent state of oscillation, so that it approximates closely to a classical vibrator, it is surely a good approximation to suppose the former to be acted upon by a prescribed force.

Parametric excitation

We have already noted that Ehrenfest's theorem leads to classical behaviour **15** for $\langle x \rangle$ under the influence of parametric excitation, and the classical ensemble just developed may be used in this case also to illustrate how $\mathscr{P}(\xi, \tau)$ develops during parametric excitation. Rather than giving a complete demonstration we shall be content with a single example, worked through in detail to show the essential characteristics of the quantal behaviour. Let the vibrator be initially in its (unnormalized) ground state, $\Psi_0 = \mathrm{e}^{-\frac{1}{2}(\xi^2 + \mathrm{i}\tau)}$, and let us seek a solution of (43) of the form

$$\Psi = \mathrm{e}^{-(y\xi^2 + k)}, \tag{51}$$

y and k both being functions of τ. For this solution to apply the following conditions must hold:

$$k = \mathrm{i}\int^{\tau} y\,\mathrm{d}\tau, \tag{52}$$

and

$$2\mathrm{i}\,\mathrm{d}y|\mathrm{d}\tau - 4y^2 + 1 + \beta\cos 2r\tau = 0. \tag{53}$$

The standard procedure[6] for an equation of Riccati type, such as (53), is to introduce $u(\tau)$, such that $2\mathrm{i}uy = \mathrm{d}u|\mathrm{d}\tau$. Then

$$\mathrm{d}^2u|\mathrm{d}\tau^2 + (1+\beta\cos 2r\tau)u = 0, \tag{54}$$

which is just the classical equation of motion for a parametrically excited vibrator. The analysis in chapter 10 shows that if the excitation is weak ($\beta \ll 1$) and perfectly in tune ($r = 1$), an adequate solution of (54) takes the form:

$$u = \mathrm{e}^{\Delta_c\tau}\cos(\tau + \tfrac{1}{4}\pi) + C\,\mathrm{e}^{-\Delta_c\tau}\cos(\tau - \tfrac{1}{4}\pi) \tag{55}$$

I.294

in which $\Delta_c = \frac{1}{4}\beta$. This case of perfect tuning is enough to illustrate the behaviour and we shall not go beyond it. The corresponding form of y follows immediately:

$$y(\tau) \doteqdot \tfrac{1}{2}\mathrm{i}[\mathrm{e}^{\Delta_c\tau}\sin(\tau+\tfrac{1}{4}\pi) + C\,\mathrm{e}^{-\Delta_c\tau}\sin(\tau-\tfrac{1}{4}\pi)]/[\mathrm{e}^{\Delta_c\tau}\cos(\tau+\tfrac{1}{4}\pi) + C\,\mathrm{e}^{-\Delta_c\tau}\cos(\tau-\tfrac{1}{4}\pi)].$$

In writing this we have assumed the excitation to be so weak ($\Delta_c \ll 1$) that the amplitude does not change appreciably in one cycle, and have accordingly ignored terms arising from derivatives of the exponential factors. If C is assigned the value i, $y(0)$ takes the ground state value $\frac{1}{2}$, the appropriate starting condition if the excitation is switched on when $\tau = 0$. Subsequently y oscillates at twice the natural frequency and the wave-function (51), while retaining its Gaussian profile, oscillates in width. The function $k(\tau)$ is a rounded staircase function, becoming squarer as time goes on, but fortunately there is no need to evaluate it. Its imaginary part contributes phase changes across the wave-function that disappear from $\mathcal{P}(\xi, \tau)$, while its real part serves to ensure conservation of $\int \mathcal{P}\,\mathrm{d}\xi$ throughout the oscillations of width. In this knowledge we write

$$\mathcal{P}(\xi,\tau) \propto (\overline{\xi^2})^{-\frac{1}{2}}\,\mathrm{e}^{-\xi^2/2\overline{\xi^2}}, \tag{56}$$

where

$$\overline{\xi^2} = 1/\mathrm{Re}[4y] = \tfrac{1}{2}\left(p^2\cos^2\theta + \frac{1}{p^2}\sin^2\theta\right), \tag{57}$$

$$p = \mathrm{e}^{\frac{1}{4}\beta\tau} \quad \text{and} \quad \theta = \tau - \tfrac{1}{4}\pi.$$

As p grows exponentially with time the pulsations of width become ever stronger, and the wave-packet oscillates vigorously twice a cycle between p and $1/p$ times its initial width.

This is what the classical ensemble predicts for the same excitation applied to all its members. If the vector representing any one classical vibrator is resolved at time $\tau = 0$, when the excitation is applied, into orthogonal components at $\pm\frac{1}{4}\pi$ to the real axis, each component develops independently, one growing exponentially as $e^{\frac{1}{4}\beta\tau}$ and the other decaying as $e^{-\frac{1}{4}\beta\tau}$. An observer rotating at the natural frequency would see the same scale transformation applied to each vector, its component at $-\frac{1}{4}\pi$ being increased and that at $\frac{1}{4}\pi$ decreased by a factor p. The initial isotropic σ, of the form $e^{-(\xi^2+\kappa^2)}$, becomes distorted into an elliptical Gaussian distribution, as in fig. 3, with axial ratio p^2, and in the laboratory framework this ellipse sweeps round at the natural frequency, its projection on the ξ-axis generating $\mathscr{P}(\xi, \tau)$ in accordance with (56), including the pre-exponential factor.

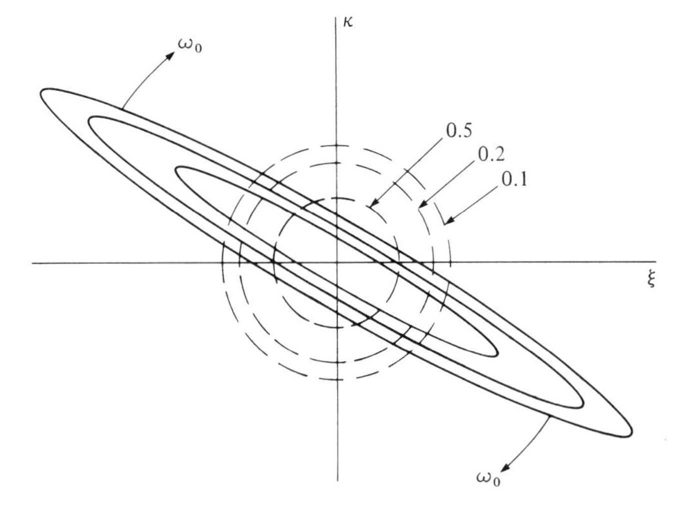

Fig. 3. The ground state, shown in the σ-representation as a circular Gaussian distribution (broken contours), is deformed by parametric excitation into an elliptical Gaussian distribution (full contours). In this case $p = 3$ and the axial ratio is 9.

It will be observed that $\mathscr{P}(\xi)$ remains even at all times in this case, and that $\langle x \rangle = 0$. This must be true, by symmetry, whenever the parametrically excited vibrator starts in a stationary state, and of course is also true for a classical vibrator initially at rest. In the classical case there may be a latent period before the advent of some disturbance to initiate the exponential growth, but with the quantized vibrator there is no latent period for the symmetrical evolution of Ψ; even in the ground state the zero-point distribution provides something for the parametric excitation to seize on, but the resulting symmetrical oscillation of $\mathscr{P}$ has no classical analogue except for the ensemble described here.

14 Anharmonic vibrators

We begin with some examples of one-dimensional vibration in a non-parabolic potential chosen to permit complete analytical solution of Schrödinger's time-independent equation. These examples are of isochronous vibrators which classically have a frequency independent of amplitude, and which might be expected therefore to have energy levels equally spaced at a separation of $\hbar\omega_0$. This expectation turns out to be very nearly right and inspires a certain confidence in the semi-classical procedure developed by (among others) Bohr, Wilson and Sommerfeld.[1] We therefore apply this procedure to some non-isochronous systems and find once more rather good agreement with the results of exact quantum mechanics. Periodic systems in fact can often be treated semi-classically with adequate accuracy, and significant economy of effort in comparison with strict quantum-mechanical analysis. This approach pays handsomely when we turn in the next chapter to the quantization of electron cyclotron orbits which, as already discussed in chapter 8, are closely related to harmonic oscillators. Conduction electrons in semi-conductors, and still more in metals, have their behaviour modified by the lattice through which they move, and a complete quantal treatment has never been achieved. It is clear, however, from approximate calculations, often of great complexity, that the semi-classical method describes most of the interesting physical processes correctly and very simply. In chapter 15 we shall describe in outline some of the effects which can be treated quite well enough for most purposes without even writing down Schrödinger's equation.

I.237

Isochronous vibrators

It was shown in chapter 2 that the potential ax^2+b/x^2 is isochronous, with ω_0 equal to $(8a/m)^{\frac{1}{2}}$, independent of b. It might appear that when $b=0$ the frequency is $(2a/m)^{\frac{1}{2}}$, but however small b may be the potential rises to infinity at $x=0$; when $b=0$ the parabola ax^2 is bisected by an impenetrable δ-function at the origin, so that the frequency is doubled. It is when b is larger that the well becomes more symmetrical and parabolic in form, $V\sim 2(ab)^{\frac{1}{2}}+4a[x-(b/a)^{\frac{1}{4}}]^2$. To avoid the base of the well changing height with b, let us redefine V as $(Ax-B/x)^2$, where $A^2=a, B^2=b$ and the minimum value of V is zero, at $x=(B/A)^{\frac{1}{2}}$. With the same reduced coordinates as for the harmonic oscillator, $\xi=(m\omega_0/\hbar)^{\frac{1}{2}}x$ and $\varepsilon=2E/\hbar\omega_0$,

I.16

and with $\beta = B(2m)^{\frac{1}{2}}/\hbar$, Schrödinger's equation takes the form:

$$\psi'' + [\varepsilon - (\tfrac{1}{2}\xi - \beta/\xi)^2]\psi = 0. \qquad (1)$$

This can be solved in a power series once the singular behaviour at $\xi = 0$ and the asymptotic behaviour at infinity have been taken out. As for the latter, the term β/ξ becomes negligible and like the harmonic oscillator the solution that vanishes at infinity is of the form $e^{-\frac{1}{4}\xi^2}$. At the origin it is β/ξ that dominates the behaviour and, approximately,

$$\psi'' \approx \beta^2\psi/\xi^2.$$

If $\psi = \xi^s$, s being a positive number such that $s(s-1) = \beta^2$, the equation is satisfied and ψ vanishes at the origin. We therefore write

$$\psi = \xi^s\, e^{-\frac{1}{4}\xi^2} F(\xi), \qquad (2)$$

where
$$s = \tfrac{1}{2}[1 + (1 + 4\beta^2)^{\frac{1}{2}}], \qquad (3)$$

and seek for a solution $F(\xi) = \sum_0^\infty u_n \xi^n$, in which $u_0 \neq 0$. Substitution of this trial solution in (1) gives the recurrence relation:

$$\frac{u_{n+2}}{u_n} = \frac{n + s + \frac{1}{2} - \varepsilon - \beta}{(n+s+2)(n+s+1) - \beta^2} = \frac{n + s + \frac{1}{2} - \varepsilon - \beta}{(n+2)(n+2s+1)}, \qquad (4)$$

which tends to $1/n$ as $n \to \infty$. Now this asymptotic recurrence relation matches that of the series expansion of $e^{\frac{1}{2}\xi^2}$, so that if the series does not terminate, $F(\xi) \sim e^{\frac{1}{2}\xi^2}$ at large ξ; then ψ, varying as $e^{+\frac{1}{4}\xi^2}$, does not fall to zero at infinity. This is exactly the same behaviour as makes it necessary to terminate the infinite series in the solution of the harmonic oscillator problem. In this case ε must be chosen so that the numerator in (4) vanishes for some value of n; then $u_n\xi^n$ is the last term in the series, and $F(\xi)$ can grow no faster than ξ^n. The exponential $e^{-\frac{1}{4}\xi^2}$ now ensures the vanishing of ψ at infinity. Only even powers of ξ appear in $F(\xi)$, and if we write n as 2ν, the νth level has energy ε_ν given by the expression:

$$\varepsilon_\nu = 2\nu + s + \tfrac{1}{2} - \beta. \qquad (5)$$

Hence
$$E_\nu = (\nu + \gamma)\hbar\omega_0, \qquad (6)$$

where, from (3),
$$\gamma = \tfrac{1}{2}[1 - \beta + (\beta^2 + \tfrac{1}{4})^{\frac{1}{2}}]. \qquad (7)$$

The energy levels are exactly spaced at $\hbar\omega_0$, as with the harmonic oscillator, but the fractional correction, γ, determining the zero-point energy $\gamma\hbar\omega_0$, is not $\frac{1}{2}$ except when β is very large and the well approximates to a parabolic potential. When $\beta = 0$, $\gamma = \frac{3}{4}$ as would be expected for a parabola bisected by a vertical barrier. For if the barrier were removed the natural frequency of the resulting harmonic oscillator would be $\frac{1}{2}\omega_0$ and the energy levels would be $(n + \frac{1}{2})\frac{1}{2}\hbar\omega_0$; however, only odd values of n generate wave-functions that vanish at the origin and hence can be fitted into the bisected potential. Writing $n = 2\nu + 1$ gives energy levels $(\nu + \frac{3}{4})\hbar\omega_0$, as required by (7). Other values of β span the range $\frac{1}{2}$ to $\frac{3}{4}$ for γ, but otherwise leave the level structure unaltered. Of course, the wave-function itself is different from that for a harmonic oscillator, being necessarily asymmetric to conform to the shape of the potential well, but perhaps less distorted than one might have guessed. Fig. 1 presents a comparison for the case $\nu = 3$, the horizontal scales having been adjusted so as to make the classical turning-points coincide.

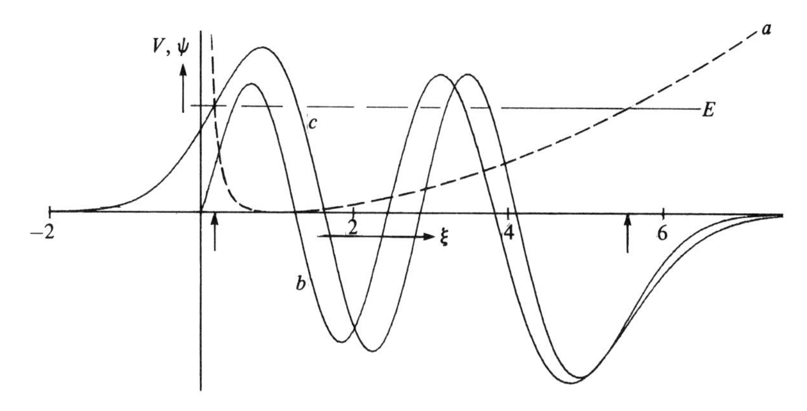

Fig. 1. The stationary state wave-function ψ_3 (curve b) for the isochronous potential $(\frac{1}{2}\xi - \beta/\xi)^2$, with $\beta = \frac{1}{2}$ (curve a). The arrows show the classical turning-points, $V = E$, and curve c is the corresponding solution for a harmonic vibrator having the same turning points.

To avoid any suspicion that all isochronous vibrators have strictly even level spacings, let us analyse a counter-example in which the potential is formed from two parabolic half-wells joined at their lowest points:

$$\left.\begin{aligned} V &= \tfrac{1}{2}m\omega_0^2x^2 \quad \text{for } x > 0 \\ &= \tfrac{1}{2}\beta^2 m\omega_0^2 x^2 \quad \text{for } x < 0, \end{aligned}\right\} \tag{8}$$

leading to a frequency $2\beta\omega_0/(1+\beta)$. On either side of the origin wave-functions that vanish at great distances can be found for all values of E, and quantization now arises from the necessity of making ψ and ψ' continuous at the origin. Let us first write down the general solution for $x > 0$, for which (13.2) applies with the same choice of reduced variables. Writing ψ as $e^{-\frac{1}{2}\xi^2}\sum_n u_n\xi^n$ we have the recursion formula

$$\frac{u_{n+2}}{u_n} = \frac{2n+1-\varepsilon}{(n+1)(n+2)}, \tag{9}$$

but we are not at liberty to choose ε so that the series terminates. Instead we recognize that (9) generates two independent series, one of even and the other of odd terms only, and that for each, since $u_{n+2}/u_n \to -2/n$ as $n \to \infty$, the behaviour at large ξ is as e^{ξ^2}. To make the solution acceptable we must arrange that the two series cancel each other at large ξ. Now the gamma function, $\Gamma(z)$, has the property[2] that for any z, $\Gamma(z+1) = z\Gamma(z)$, and this enables the coefficients u_n to be expressed compactly:

If n is even $\qquad u_n = 2^n\dfrac{\Gamma(\frac{1}{2}n - y)}{\Gamma(-y)n!}u_0;$

and
$$u_{n+1}=2^n\frac{\Gamma[\frac{1}{2}(n+1)-y]}{\Gamma(-y+\frac{1}{2})(n+1)!}u_1,$$

where $y=\frac{1}{4}(\varepsilon-1)$. The terms in each series take the same form if $u_1/u_0=2\Gamma(-y+\frac{1}{2})/\Gamma(-y)$, and they cancel each other asymptotically if the signs are opposite, i.e.

$$u_1/u_0=-2\Gamma(-y+\tfrac{1}{2})/\Gamma(-y)=2y\tan\pi y\cdot\Gamma(y)/\Gamma(y+\tfrac{1}{2}).$$

Now at the origin $(\mathrm{d}\psi/\mathrm{d}\xi)/\psi=u_1/u_0$, and therefore in laboratory coordinates

$$(\psi'/\psi)_0=\tfrac{1}{2}(m\omega_0/\hbar)^{\frac{1}{2}}(\varepsilon-1)\tan[\tfrac{1}{4}\pi(\varepsilon-1)]\Gamma[\tfrac{1}{4}(\varepsilon-1)]/\Gamma[\tfrac{1}{4}(\varepsilon+1)]. \quad (10)$$

This is shown as curve 1 in fig. 2; the curve crosses the axis at those values of E which are even stationary states ($\psi'=0$) of the harmonic oscillator, and rises to ∞ at the odd stationary states ($\psi=0$).

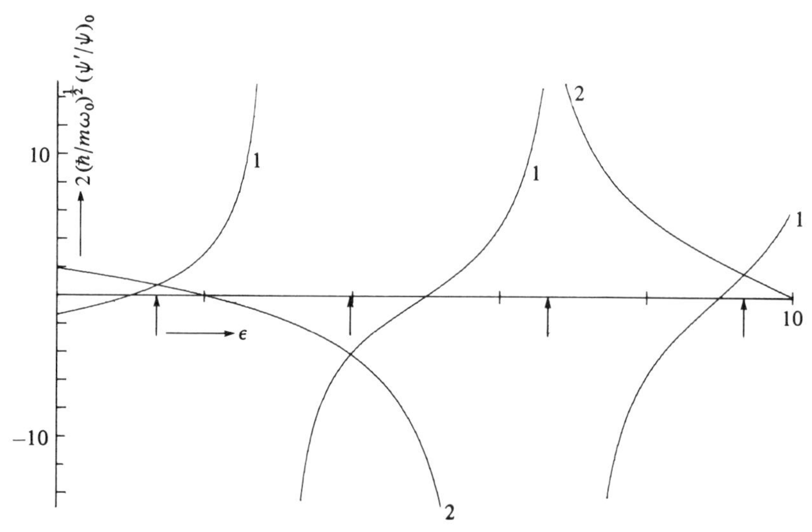

Fig. 2. Illustrating the energy levels for a particle in an asymmetric parabolic potential as the intersections (shown by arrows) of curves 1 and 2; see text for a full description.

On the negative side, $x<0$, the same behaviour holds except that the sign of (ψ'/ψ) is reversed and ω_0 is replaced by $\beta\omega_0$. Thus to apply to the same values of E curve 2 must be derived from curve 1 by scaling horizontally by β, vertically by $\beta^{\frac{1}{2}}$, and reflecting in the horizontal axis. In this case β has been taken as 2. The intersections give those values of $E/\hbar\omega_0$ at which ψ and ψ' may be matched at the origin, and are therefore the stationary energy states.

Since the frequency is $\frac{4}{3}\omega_0$, one might expect eigenvalues of $(n+\frac{1}{2})\frac{4}{3}\hbar\omega_0$, and this is very nearly realized. When the energies are expressed as $(n+\gamma_n)\frac{4}{3}\hbar\omega_0$, computed values of the intersections give $\gamma_0=0.50895$, $\gamma_1=0.49497$, $\gamma_2=0.49937$, $\gamma_3=0.49890$, $\gamma_4=0.50012$ etc. Departures from strictly even spacing of the levels undoubtedly occur, but they are extremely small. The somewhat capricious deviations of γ from $\frac{1}{2}$ are probably attributable to the discontinuity in $\mathrm{d}^2V/\mathrm{d}x^2$ at $x=0$, whose effect will depend on whether ψ is large or small there. We shall see similar examples presently.

Arbitrary potential well; the semi-classical method

The potentials just analysed happen to submit to analytical treatment without difficulty, but an arbitrarily chosen form of $V(x)$ is unlikely to yield to anything but a computer. If $V(x)$ is symmetrical about the origin the fact that ψ is either odd or even simplifies the computation, for one can start a trial integration of Schrödinger's equation at the origin with either ψ or ψ' equal to zero, and seek a value of E that causes convergence to zero at a great distance. In the absence of symmetry the search is more tedious, for in addition to E the value of ψ'/ψ at some point must be correctly adjusted. This process may be obviated if only the energy levels are sought, without inordinate accuracy, and not the actual wave-functions. It may be adequate to assume the spacing of neighbouring energy levels to be $\hbar\omega$, making allowance for non-isochronous potentials by the appropriate variation of ω with E. This procedure derives essentially from the old quantum mechanics associated with the names of Bohr, Wilson and Sommerfeld, to which should be added the name of Ehrenfest who stressed that *adiabatic invariants* provide a logical underpinning for the rules of quantization that others had discovered heuristically. It is worth devoting some space to this matter, old-fashioned though it may be, since we shall find on comparing its predictions with the correct answers, in a few cases for which the latter can be found without special computation, that the semi-classical recipe is often remarkably good. The standard general treatment, based on Hamiltonian dynamics, is accessible elsewhere[3] and here we shall derive by elementary means the particular form needed for anharmonic vibrators; this will serve to give physical substance to what is apt to appear a dry formalism, somewhat obscure in meaning.

When a particle vibrates freely, and without dissipation, in a static potential $V(q)$, its coordinate q and momentum p vary periodically so that they may be represented by a closed curve on the q–p phase plane. A harmonic vibrator, with q and p varying sinusoidally in phase quadrature, executes elliptical trajectories, but in general all we can say is that there will be a family of closed curves, one for each energy of vibration, as in fig. 3, with mirror symmetry about the q-axis since reversing the sign of p leaves E unchanged. We now imagine the potential V to be very slowly deformed, allowing many cycles of oscillation for any appreciable change to occur. Any quantity which is unchanged by this process is an adiabatic invariant. The total energy is certainly not an adiabatic invariant, since at any instant the raising of V at the point where the particle happens to be increases its potential energy without immediately affecting its kinetic energy; i.e. $\dot{E}=\dot{V}$, in which $\dot{V}$ must be understood as the rate of change of V at the location of the particle. The change of E and the deformation of

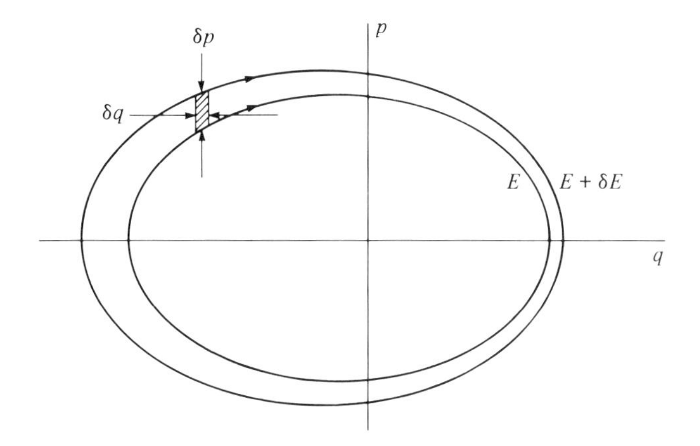

Fig. 3. Phase-plane trajectories for a classical anharmonic vibrator.

V both result in modifications of the shape of the phase-trajectory, but its area J, which may also be written $\oint p\,\mathrm{d}q$, remains invariant. To prove this result, let us consider the process in two stages – first we slowly make a small deformation, $\delta V(q)$, to V and calculate the resulting change in area, assuming E to remain constant; then we add the effect of the change in E that accompanies $\delta V(q)$. For the first stage, note that if (q, p) is a point on the original trajectory, where the velocity p/m is v, then $(q, p - \delta V(q)/v)$ is a point on the final trajectory of the same energy; the increase $\delta V(q)$ in potential energy must be compensated by a decrease in kinetic energy, resulting in a decrease of momentum δp, such that $v\delta p = \delta V$. In this process the area of the cycle is diminished by $\oint \delta p \cdot \mathrm{d}q$,

i.e. $$\delta J_1 = -\oint \delta V \cdot \mathrm{d}q/v.$$

For the second stage, note that if T is the periodic time of the vibration, the time $\mathrm{d}q/v$ spent by the particle in the element $\mathrm{d}q$ is a fraction $\mathrm{d}q/vT$ of the cycle; it therefore picks up this fraction of the change in V as it passes through $\mathrm{d}q$. Overall, $\delta E = \oint \delta V \cdot \mathrm{d}q/vT$ and we have for the corresponding change in area,

$$\delta J_2 = (\oint \delta V \cdot \mathrm{d}q/vT)\,\mathrm{d}J/\mathrm{d}E.$$

If we can show that $\mathrm{d}J/\mathrm{d}E = T$ it will follow that $\delta J_1 + \delta J_2 = 0$ and J is invariant under slow deformation of V. To achieve this final stage, consider two trajectories δE apart and imagine the representative point as it traverses the cycle sweeping out the area δJ of the enclosed annulus. Since a typical element of area is $\delta p\mathrm{d}q$, as shown in fig. 3, the rate of sweeping out of the annular area, $\mathrm{d}(\delta J)/\mathrm{d}t$, is $v\delta p$ which is constant and equal to δE; hence, integrating round the whole cycle, we arrive at the required result, $\delta J/T = \delta E$ or, in the limit,

$$\mathrm{d}J/\mathrm{d}E = T. \tag{11}$$

Consequently, J is an adiabatic invariant.

The use of this result to generalize Bohr's quantum mechanics stemmed from the plausible belief that if a system in an acceptable quantum state is very slowly deformed there can never be a moment at which it is stimulated to jump to another state; the slow deformation must therefore take it automatically through a continuous sequence of acceptable states. Each quantum state is defined by assigning one or more of a discrete set of quantum numbers to it, these numbers serving to characterize the dynamical state; thus in the quantization of harmonic oscillators we know that the quantity $E/\hbar\omega_0$ takes one of the half-integral values $n+\frac{1}{2}$, and in Bohr's quantization of the hydrogen atom the only allowed values of angular momentum L are such that $L/\hbar = n$. If there is to be no problem about the system needing to jump to another state during slow deformation, the adiabatic invariants, of which J is an example, are ideal choices of physical variables to which quantum numbers may be assigned. The two examples just mentioned illustrate this point. A harmonic oscillator has an elliptical phase trajectory with semi-axes q_0 (the amplitude) and $m\omega_0 q_0$; its area J is $\pi m\omega_0 q_0^2$, which is $2\pi E/\omega_0$. Quantization of J as $(n+\frac{1}{2})h$ quantizes E as $(n+\frac{1}{2})h\omega_0/2\pi$. As for an electron in a central orbit, the angular momentum is invariant under changes of the force law and is therefore a natural candidate for quantization. For the moment, however, we restrict the discussion to systems of one variable, for which the quantization rule of Sommerfeld and Wilson may be written:

$$J = \oint p \, \mathrm{d}q = (n+\gamma)2\pi\hbar, \tag{12}$$

the fractional correction γ being a later refinement.

For an isochronous vibrator $T = 2\pi/\omega_0$ and is constant, so that (11) gives J as $2\pi E/\omega_0$ and Planck's quantum law follows from (12) in the form $E = (n+\gamma)\hbar\omega_0$. Although the determination of γ is beyond the range of semi-classical quantum mechanics, it is a tribute to the power of the method that for the whole set of isochronous potentials in (1) γ was found to be constant for each choice of β, and for the whole range of β only varied between $\frac{1}{2}$ and $\frac{3}{4}$. Let us test it further by determining γ and its variation with energy for some non-isochronous potentials where the true answer can be found by reference to tabulated functions. First, the square well, with V equal to zero when $-\frac{1}{2}a < x < \frac{1}{2}a$ and infinite otherwise, so that the classical particle bounces back and forth at constant speed between hard walls. The phase trajectory is a rectangle of width $\Delta q = a$, and height $\Delta p = 2(2mE)^{\frac{1}{2}}$; hence $J = 2a(2mE)^{\frac{1}{2}}$. The energy of the nth level then follows from (12):

$$E_n = (n+\gamma)^2\pi^2\hbar^2/2ma^2, \tag{13}$$

which is to be compared with the correct result, $(n+1)^2\pi^2\hbar^2/2ma^2$, resulting from the requirement that the de Broglie wavelength shall be $2a/(n+1)$. In this case $\gamma = 1$, independent of energy. It may be noted that when $\beta = 0$ in (1), and the potential is parabolic on one side and hard on the other, the true value of γ, $\frac{3}{4}$, is the mean of $\frac{1}{2}$ for the former and 1 for the latter. It will become apparent in the following discussion of the WKB method that γ represents a correction arising from the boundary conditions satisfied by ψ at the ends of the trajectory, and that the value for the complete oscillation is always an arithmetic mean of contributions from the two ends. In saying this, we ignore the weak effect noted earlier, where slight fluctuations in γ were attributed to lack of smoothness of the potential.

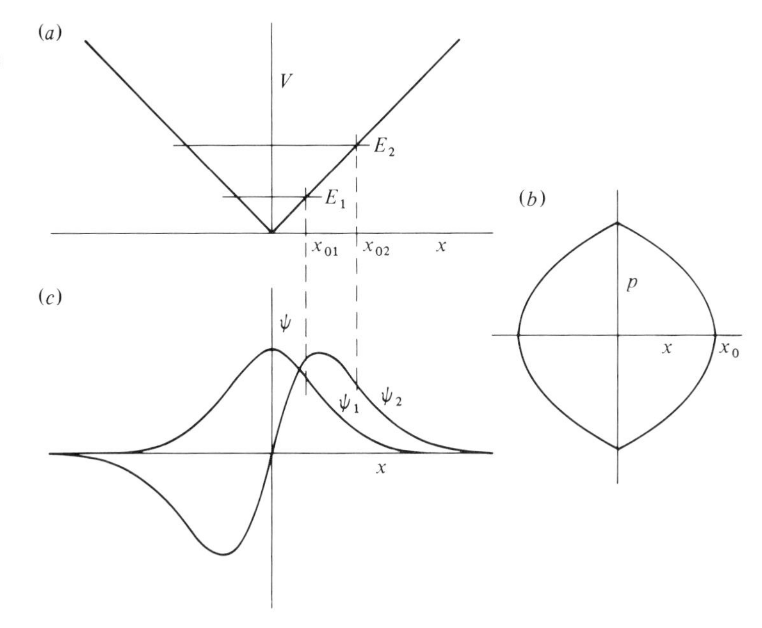

Fig. 4. A particle vibrating in the potential (a) has the classical phase-plane trajectory (b); the lowest two solutions of Schrödinger's equation are shown in (c), with the classical turning points indicated by vertical broken lines.

Next let us consider a symmetrical potential well with sloping sides, $V = \alpha|x|$, as in fig. 4(a). The classical particle experiences the same steady acceleration back towards the origin whichever way it is displaced. If the amplitude is x_0, $p^2 = 2m\alpha[x_0 - |x|]$, giving phase trajectories in the form of two parabolic arcs, as in fig. 4(b). The area is $\frac{8}{3}x_0^{\frac{3}{2}}\sqrt{2m\alpha}$, and therefore

$$J_n = (8/3\alpha)(2mE_n^3)^{\frac{1}{2}} = (n+\gamma)2\pi\hbar, \text{ from (12);}$$

consequently,
$$E_n = [\tfrac{3}{4}\pi\alpha\hbar(n+\gamma)]^{\frac{2}{3}}/(2m)^{\frac{1}{3}}. \tag{14}$$

To compare this with the true solution we note that V is symmetrical about the origin and that the solutions will have either $\psi(0)$ or $\psi'(0)$ equal to zero. It is necessary therefore to consider only solutions for positive x, with Schrödinger's equation in the form:

$$\hbar^2\psi''/2m + (E - \alpha x)\psi = 0; \qquad x > 0. \tag{15}$$

Now move the origin to the classical limit, $x_0 = E/\alpha$, as shown in fig. 4(a), and write z for $(2m\alpha/\hbar^2)^{\frac{1}{3}}(x - x_0)$. Then

$$\mathrm{d}^2\psi/\mathrm{d}z^2 - z\psi = 0; \qquad z > -(2m/\hbar^2\alpha^2)^{\frac{1}{3}}E. \tag{16}$$

This is Airy's equation, whose solutions are extensively tabulated.[4] In the present case we need only note that the physically sensible solution is $\mathrm{Ai}(z)$, which decays smoothly to zero when $z>0$ and is oscillatory for $z<0$. The quantized energies are those that cause $\mathrm{Ai}(z)$ or its derivative to vanish when $-z=z_n=(2m/\hbar^2\alpha^2)^{\frac{1}{3}}E_n$; then the complete wave-function, of which fig. 4(c) shows examples, is constructed from this function and its mirror image, with reversed sign if necessary so that ψ and ψ' are continuous at the origin of x. The zeros of Ai and Ai′ alternate, and are tabulated to a degree of accuracy far exceeding what is needed here.[5] These provide the allowed values of z_n, from which we can discover without further computation the values of $E_n=(\hbar^2\alpha^2/2m)^{\frac{1}{3}}z_n$. Comparison with (14) yields the values of γ needed to produce the right answer by the semi-classical process:

$$n+\gamma=4z_n^{\frac{3}{2}}/3\pi,$$

and with $z_n=1.0188$, 2.3381, 3.2482, 4.0879 etc., we find the following sequence: $\gamma_0=0.436$, $\gamma_1=0.517$, $\gamma_2=0.485$, $\gamma_3=0.508$, $\gamma_4=0.491$, $\gamma_5=0.505,\ldots,\gamma_{99}=0.4996$, $\gamma_{100}=0.5003$ etc.

After only a few terms γ has settled down to values very close to $\frac{1}{2}$, though there remains a slight alternation about the mean which is attributable to the cusp in V at the centre. To a very small degree it matters whether ψ has a maximum or is zero at the cusp.

For a third example consider the Morse potential, $V_0(1-\mathrm{e}^{-a(r-r_0)^2})$, which has frequently been used to model the interaction between the atoms in a diatomic molecule whose binding energy is V_0 and whose equilibrium separation is r_0. With this potential Schrödinger's equation can be solved in series to yield a simple closed form for the energy levels[6]:

$$E_n=(n+\tfrac{1}{2})\hbar\omega_0-\tfrac{1}{4}(n+\tfrac{1}{2})^2\hbar^2\omega_0^2/V_0, \qquad (17)$$

in which ω_0 is $a(2V_0/m)^{\frac{1}{2}}$, the frequency for small-amplitude vibrations of the corresponding classical vibrator. The general expression for the variation of frequency with energy is readily derived by integrating (2.11): $\omega(\varepsilon)=\omega_0(1-\varepsilon)^{\frac{1}{2}}$, where $\varepsilon=E/V_0$, and the phase integral is equally readily evaluated:

$$\oint p\,\mathrm{d}q=(2\pi\omega_0 m/a^2)[1-(1-\varepsilon)^{\frac{1}{2}}].$$

Semi-classical quantization by use of (12) agrees perfectly with (17) of $\gamma=\frac{1}{2}$.

These examples suffice to illustrate the merit of the semi-classical approach, and if we now proceed to yet another it is more to draw attention to the perturbation technique which is of value when the form of V is close to one for which Schrödinger's equation is soluble in terms of well-understood functions. The mathematical procedure is described in standard texts,[7] and we need only quote a typical result, choosing a symmetrical, slightly anharmonic, potential in which the quadratic term is supplemented by a quartic:

$$V = \tfrac{1}{2}m\omega_0^2x^2 + \beta x^4. \quad (18)$$

The perturbation approach, which expresses ψ as a Fourier series, using the harmonic oscillator wave-functions as basis, yields the energy levels as a power series in β. We shall assume β small enough to allow the series to be terminated after the first-order correction; then

$$E_n \approx \hbar\omega_0[(n+\tfrac{1}{2}) + (n^2+n+\tfrac{1}{2})\Delta], \quad (19)$$

where $\Delta = 3\beta\hbar/2m^2\omega_0^3$. Let us see what value γ must take in the semi-classical approach to give the same answer. If $\beta = 0$, the trajectory for each E is an ellipse of area $2\pi E/\omega_0$, and the first-order correction due to β is easily calculated. At any value of p, (18) shows that x must be reduced by δx, equal to $\beta x^3/m\omega_0^2$, if V and hence E are to be unchanged. The resulting reduction in area is $\oint \delta x \cdot \mathrm{d}p$, and for the first-order correction p can be written in its unperturbed form, $[2m(E - \tfrac{1}{2}m\omega_0^2x^2)]^{\frac{1}{2}}$. Hence we find that

$$J = 2\pi E/\omega_0 - 3\pi\beta E^2/m^2\omega_0^5 = 2\pi E(1 - E\Delta/\hbar\omega_0)/\omega_0.$$

If E_n is given by (19),

$$J_n \approx (n + \tfrac{1}{2} + \tfrac{1}{4}\Delta)2\pi\hbar, \text{ to first order in } \Delta,$$

which accords with (12) if $\gamma = \frac{1}{2} + \frac{1}{4}\Delta$, independent of n. The semi-classical solution is perfectly correct as far as it goes – the true energy levels (at least to first order in β) correspond to exactly equal increments of J; since there is no pretence of calculating γ, the solution cannot be faulted for not predicting the extent to which it differs from $\frac{1}{2}$. It will be observed that the correction to γ is positive, as might be expected from the increased steepness of the potential resulting from the quartic terms. To obtain some idea of the magnitude of the correction, let us suppose that the quartic term is large enough to raise the 5th level by $\hbar\omega_0$, i.e. in (19) let $(n^2+n+\frac{1}{2})\Delta$ be unity when $n = 5$. This is a substantial anharmonicity, but Δ is still only 3.3×10^{-2}, and $\gamma = 0.508$, very little affected in fact.

WKB approximation

In almost all these tests the semi-classical approach, with γ put equal to $\frac{1}{2}$, has shown itself remarkably successful in predicting the energy levels, but of course it does not yield the wave-functions and is therefore powerless for any other quantum-mechanical calculation such as the expectation values of physical quantities or transition probabilities under the influence of time-dependent perturbations. Direct integration by computer of Schrödinger's equation in one dimension presents no problems and is always available if a certain numerical result is all that is wanted. If, however, an analytical form is desired it is worth bearing in mind that the WKB approach[8] offers an intermediate stage of approximation which may be good enough, especially for high quantum numbers. In its most elementary form, which is the form that is most generally useful, the solution of the equation

$$\psi'' + \psi f(x) = 0$$

is written as

$$\psi \approx Cf^{-\frac{1}{4}} \exp\left[i \int_{x_0}^{x} f^{\frac{1}{2}}\, dx\right] \tag{20}$$

This result is not peculiar to wave mechanics, but is a useful approximate form that applies to any problem of wave propagation along a line where the properties of the medium vary with x; $f^{\frac{1}{2}}$ is the local value of the wavenumber, and $\int_{x_0}^{x} f^{\frac{1}{2}}\, dx$ is the phase difference between x_0 and x, on the assumption that the local wavenumber, in spite of varying with x, still describes the local rate of change of phase with x. Since the phase and group velocities of the wave change with x one must expect the amplitude also to change in order to conserve the flux of whatever property is being transported by the wave, and the coefficient $f^{-\frac{1}{4}}$ takes care of this.† The solution may conveniently be illustrated and checked by reference to Airy's equation (16), for which $f(x) = -x$ as in fig. 4(a), and for which (20) may be written in the form

$$\psi = C\bar{x}^{-\frac{1}{4}} \exp\left(\tfrac{2}{3} i \bar{x}^{\frac{3}{2}}\right). \tag{21}$$

Here x_0 is incorporated in the complex constant C, and $\bar{x}$ is written for $-x$. This describes the oscillatory solution for negative x, the region in which $V < E$ and a classical particle can be found. As written, (21) describes a travelling wave accompanying the particle as it descends the potential slope, but for a particle bound in a V-shaped potential a standing wave is appropriate, as described by Re $[\psi]$. Whatever the phase of C, Re $[\psi]$ is a sinusoid of varying wavelength and with an amplitude enclosed within an envelope that varies as $\bar{x}^{-\frac{1}{4}}$; but the phase of C determines the placing of the zeros. This raises a question that cannot be answered within the framework of the elementary WKB theory. It is necessary to continue the solution through the point $x = 0$ into the classically inaccessible region where the wave-function must decay rapidly and vanish, and this decay occurs only if the zeros are correctly placed. The mathematical analysis of this problem by development of asymptotic expansions for the solution of (16) is subtle and historically important, but now that extensive tabulations of Airy functions are available we may appreciate the physical point merely by looking at graphs. Fig. 5 shows the physically valid function, Ai (x), which vanishes at $+\infty$, and for comparison the WKB solutions on either side of the origin, C being chosen for the latter so as to make the match

† $\psi\psi^* \propto f^{-\frac{1}{2}}$, and since $f = 2m(E-V)/\hbar^2$, $f^{\frac{1}{2}}$ is proportional to the velocity of a classical particle at x. The WKB solution thus reproduces in its amplitude variations the classical expectation of finding the particle at any point.

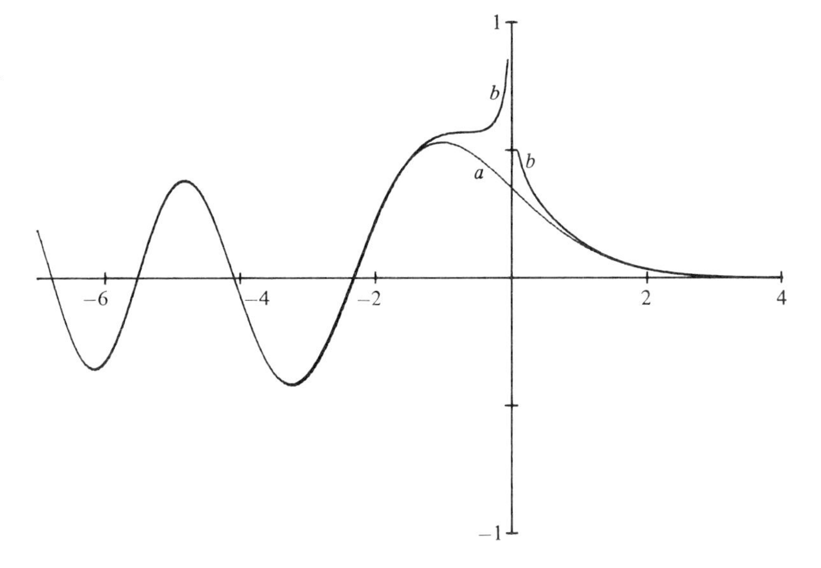

Fig. 5. Solution of Airy's equation: (*a*) exact, (*b*) WKB approximation fitted asymptotically on both sides.

as good as possible far from the origin:

$$\begin{aligned} \psi_{\mathrm{WKB}} &= \pi^{-\frac{1}{2}}\bar{x}^{-\frac{1}{4}} \sin\left(\tfrac{2}{3}\bar{x}^{\frac{3}{2}} + \tfrac{1}{4}\pi\right); \qquad x<0 \\ &= \tfrac{1}{2}\pi^{-\frac{1}{2}} x^{-\frac{1}{4}} \exp\left(-\tfrac{2}{3}x^{\frac{3}{2}}\right); \qquad x>0. \end{aligned} \tag{22}$$

These expressions are in fact just the first terms in the asymptotic expansions of Ai (x).

It is clear from the diagram that the WKB solution is extremely good except near the origin, where the local wavenumber (or extinction coefficient for positive x) suffers a relatively large change in the course of a wavelength. The analysis has shown that to a very good degree of approximation one may calculate the energy levels by treating the de Broglie wave as a wave on a resonant line, using the local wavenumber to determine the phase length and adding an eighth of a wavelength to each end (the $\frac{1}{4}\pi$ in the first equation of (22)) as end corrections. This prescription is indeed exactly the semi-classical quantization process, the value of $\frac{1}{2}$ assigned to γ being this end correction. We have already seen from the tabulated values of γ how good the prescription is, and now in fig. 5 we see how to proceed beyond this point and obtain by the WKB process a generally good picture of the wave-function itself.

All this, of course, has been developed by reference to one equation only, but there is no difficulty in believing that different smooth variations of V will be equally well served by the WKB approximation. Only at the classical turning point, where $E=V$, does it break down completely, and then it may well be possible to treat V as varying linearly with x through this critical range, so that the WKB solutions on either side are matched to the Airy function in the range itself. For this to work satisfactorily it is only necessary that V shall be nearly linear for a distance covering at least one cycle of oscillations. When the condition is satisfied the semi-classical prescription works well and γ may be taken as $\frac{1}{2}$. For a more severe test we may consider a flat-bottomed potential well whose ends rise linearly, as in fig. 6; by varying the slope we may follow the change of γ from $\frac{1}{2}$

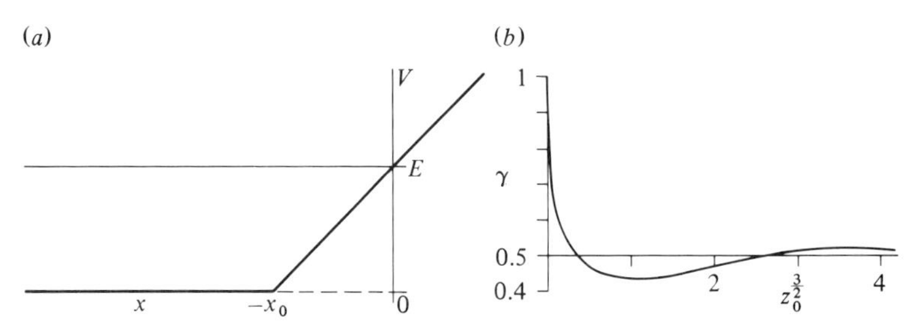

Fig. 6. Illustrating the end-correction γ for a flat-bottomed potential well with sloping ends (curve a); $z_0^{\frac{3}{2}}$ in curve b is the phase length of the sloping section of V between $-x_0$ and the classical turning point, 0.

when it is a gentle gradient to 1 when it is a vertical wall. Only one end need be considered, and if the origin is taken at the classical limit it is required to match the Airy function that applies when $x > -x_0$ to the sinusoidal standing wave that applies when $x < -x_0$. Writing α for the slope of the end, E/x_0, and z for $-(2m\alpha/\hbar^2)^{1/3}x$, we have

$$\psi = \left.\begin{array}{ll}\mathrm{Ai}\,(z); & z > -z_0 \\ C\cos(pz+\theta); & z < -z_0\end{array}\right\} \qquad (23)$$

where $z_0 = (2m\alpha/\hbar^2)^{\frac{1}{3}}x_0 \propto E^{\frac{1}{3}}x_0^{\frac{2}{3}}$, and $p = (\hbar^2/2m\alpha)^{\frac{1}{3}}(2mE)^{\frac{1}{2}}/\hbar = z_0^{\frac{1}{2}}$. In order that ψ and ψ' shall be continuous at $-z_0$,

$$z_0^{\frac{1}{2}}\tan(-z_0^{\frac{3}{2}}+\theta) = -\mathrm{Ai}'(-z_0)/\mathrm{Ai}(-z_0).$$

This equation determines θ and hence the placing of the standing wave in relation to the classical limit, $z = 0$:

$$\theta = z_0^{\frac{3}{2}} - \tan^{-1}[\mathrm{Ai}'(-z_0)/z_0^{\frac{1}{2}}\,\mathrm{Ai}(-z_0)]. \qquad (24)$$

Now the asymptotic form (22) of Ai (z) at large negative values is $\cos[\frac{1}{4}\pi - \frac{2}{3}(-z)^{\frac{3}{2}}]$, from which it follows that when the slope of V is extremely gentle,

$$\theta \approx \tfrac{1}{4}\pi + \tfrac{1}{3}z_0^{\frac{3}{2}}.$$

The term $\frac{1}{3}z_0^{\frac{3}{2}}$ is the shift of the standing wave caused by the sloping potential, relative to the position it would adopt if V remained level right up to the end, $z = 0$; that is to say, the difference between $z_0^{\frac{3}{2}}$ for constant V and $\frac{2}{3}z_0^{\frac{3}{2}}$ from (21). The first term, $\frac{1}{4}\pi$, is the phase correction we have come to expect for a linear potential variation. When z_0 is not so large the phase correction is $\theta - \frac{1}{3}z_0^{\frac{3}{2}}$. Since a phase correction of π brings the standing wave back to its original position, and since γ is defined to allow for both ends of the potential well, γ is $2/\pi$ times this phase correction:

$$\gamma = \frac{2}{\pi}\{\tfrac{2}{3}z_0^{\frac{3}{2}} - \tan^{-1}[\mathrm{Ai}'(-z_0)/z_0^{\frac{1}{2}}\,\mathrm{Ai}(-z_0)]\}. \qquad (25)$$

Tables of the Airy function[4] allow γ to be calculated very readily, with the result shown in fig. 6(b); the abscissa, $z_0^{\frac{3}{2}}$, is the phase length of the sloping section for a wave whose wavenumber is $(2mE)^{\frac{1}{2}}/\hbar$, as on the level potential, and it is at first sight rather surprising how short a sloping section ($\frac{1}{20}$ of a wavelength) is needed to bring γ down from unity to something close to $\frac{1}{2}$. The explanation lies in the ease with which the wave-function spreads into the classically forbidden region. Since the extinction coefficient depends on the square root of the energy deficit, i.e. $\psi \sim e^{-\mu x}$ where $\mu = [2m(V-E)]^{\frac{1}{2}}/\hbar$, a very steep gradient of V is needed to force ψ down so sharply that a node of the standing wave lies close to the classical limit. It is easy to show that, for small z_0, $\gamma \approx 1 - 2z_0^{\frac{1}{2}}/\pi a$, so that for a given choice of E, $(1-\gamma)$ begins to rise as $x_0^{\frac{1}{3}}$. It is evident that the WKB method, with γ fixed at $\frac{1}{2}$, is considerably more successful than one would have dared hope.

Perhaps the most important physical problems involving anharmonic vibrators arise from the spectroscopy of diatomic molecules. For small amplitudes of vibration these behave reasonably well as harmonic oscillators, and the level spacing, revealed by their spectra, is nearly constant (if we leave out of account the centrifugal stretching effects due to rotation). At higher excitations, however, the potential well is wider than required for isochronous oscillation, as shown in fig. 7, and the frequency and level

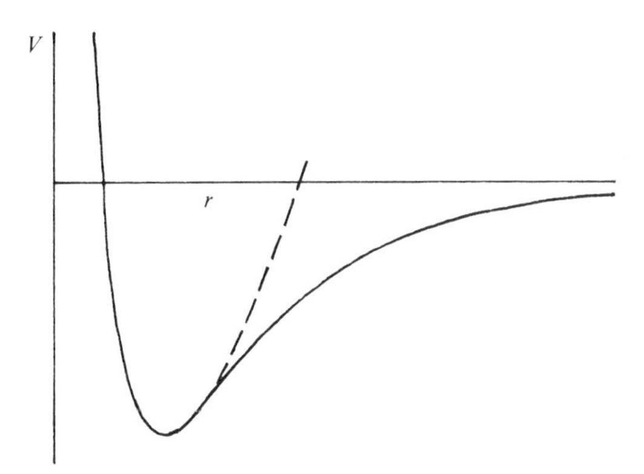

Fig. 7. Typical potential energy of interaction between two atoms (full curve) compared with an isochronous potential (broken curve).

spacing decrease. At a certain energy of excitation the molecule dissociates, and just below this the vibrational frequency is quite low. In principle it should be possible to derive a great deal of information on the shape of the potential well from the level spacing – not everything since, as with the classical oscillator, shearing of the potential curve has little or no effect on the levels. But the repulsive potential at close separations of the atoms is so steep that any plausible form for this leads to virtually the same shape at great distances, and the difficulty is not very serious. A good account of the processes available to determine $V(r)$ has been given by Le Roy,[9] and the reader will be struck by the central role played by extensions of the WKB method which, in contrast to a full wave-mechanical treatment, are manageable with fairly modest resources for computation. As a result of this work, $V(r)$ is known with considerable precision for a number of diatomic molecules; without it almost everything one could say about the details of $V(r)$ would be guesswork.

15 Vibrations and cyclotron orbits in two dimensions

For a particle moving in the quadratic potential, $V = \frac{1}{2}\mu_x x^2 + \frac{1}{2}\mu_y y^2$, Schrödinger's equation takes the form, in Cartesian co-ordinates,

$$\hbar^2(\partial^2\psi/\partial x^2 + \partial^2\psi/\partial y^2)/2m + (E - \tfrac{1}{2}\mu_x x^2 - \tfrac{1}{2}\mu_y y^2)\psi = 0, \qquad (1)$$

which is separable by writing ψ as a product, $\psi(x, y) = X(x)Y(y)$; X and Y each obey the oscillator equation (13.1):

$$\hbar^2 X''/2m + (E_x - \tfrac{1}{2}m\omega_x^2 x^2)X = 0; \qquad \hbar^2 Y''/2m + (E_y - \tfrac{1}{2}m\omega_y^2 y^2)Y = 0,$$

in which $\omega_{x,y}^2 = \mu_{x,y}/m$, and $E_x + E_y$ must equal E. Thus each direction of motion is independently quantized, and the energy levels follow the rule:

$$E = (n_x + \tfrac{1}{2})\hbar\omega_x + (n_y + \tfrac{1}{2})\hbar\omega_y. \qquad (2)$$

A typical wave-function for a stationary state, being the product of two oscillator functions, has its general form defined by a rectangular lattice of unequally spaced nodal lines, parallel to the axes, with the sign of ψ at the antinodes alternating from one cell to the next. Fig. 1(a) is a typical example in which $\omega_y = \frac{3}{4}\omega_x$, $n_x = 5$ and $n_y = 2$. The energy levels may be represented as in fig. 1(b), with each level n_x serving as the base for a ladder of n_y. Alternatively, as fig. 1(c) shows, they may be represented as a rectangular grid of points having co-ordinates $(n_x + \frac{1}{2})\hbar\omega_x$ and $(n_y + \frac{1}{2})\hbar\omega_y$; states of the same total energy lie on lines at 45°, and the linear increase in density of states with E is clearly seen in the increased length of these lines.

11, 21

The parallel between classical and quantum mechanical behaviour discussed in chapter 13 carries through from one to two dimensions, and indeed to three or more in a potential which is a positive definite quadratic form. If the system is excited from its ground state by the application of a uniform, time-dependent force, the resulting wave-function (not a stationary state, of course) takes the form of a wave-packet of the same elliptical Gaussian shape as the ground state wave-function, moving in the classical trajectory; in general this will be an evolving Lissajous figure, if ω_x and ω_y are not simply related. In the special case of an isotropic oscillator, when $\omega_x = \omega_y$, the wave-packet, which now has circular symmetry, executes a steady elliptical trajectory if left alone after excitation. Unlike a wave-packet moving in a constant potential, which gradually spreads as it travels, this particular solution of Schrödinger's equation for a harmonic potential (not necessarily isotropic) keeps its compact form indefinitely.

I.28

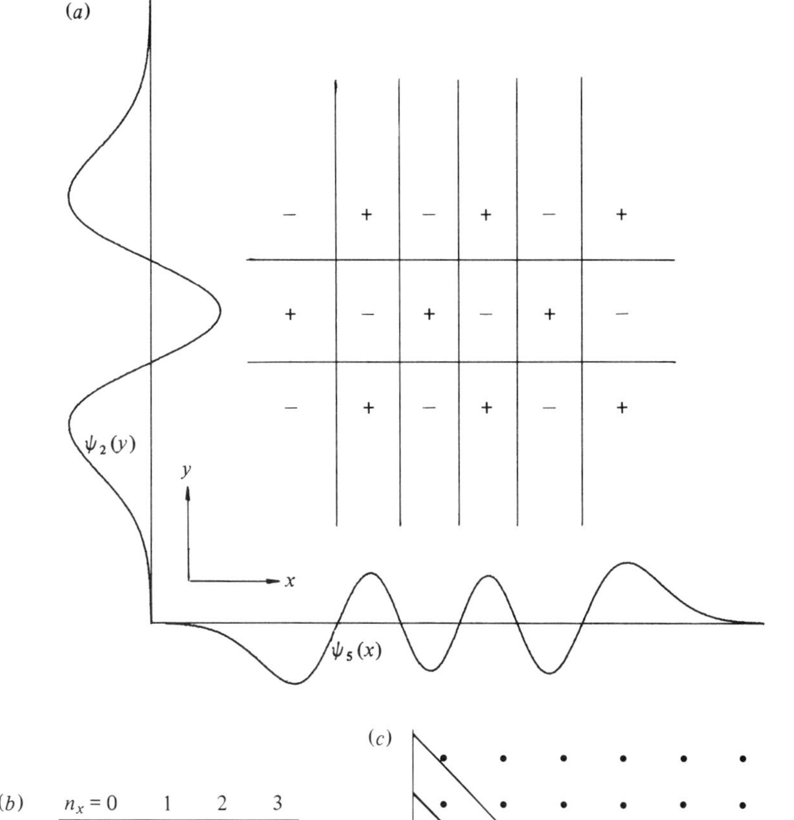

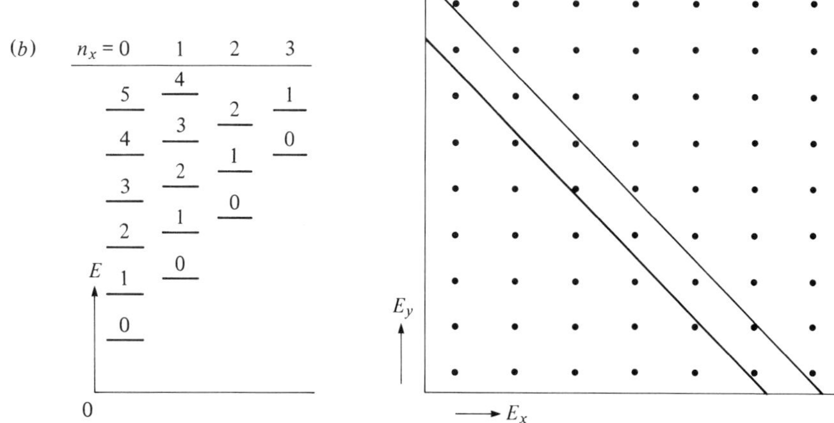

Fig. 1. (*a*) Wave-function for a two-dimensional harmonic vibrator. The stationary state is represented by a product wave-function $\psi_5(x)\psi_2(y)$, whose nodal lines form a rectangular grid; the antinodes take opposite signs in alternate cells. (*b*) Energy level diagram for the case $\omega_y = \frac{3}{4}\omega_x$; each level corresponds to a wave-function $\psi_{n_x}(x)\psi_{n_y}(y)$. Levels for each different value of n_x are shown as separate ladders, with the values of n_y attached to the levels themselves. (*c*) The same set of levels as points on a two-dimensional diagram. All points on a line at 45° have the same total energy, and the density of states is equal to the number of points between two lines corresponding to unit difference in total energy.

To look at this point from another aspect let us consider building up the wave-packet from stationary states. Since we seek illustration of the principle rather than useful results, it is sufficient to take the especially simple case of a circular trajectory, for which purpose the solutions of the equation in polar co-ordinates are appropriate. In the isotropic case $V = \frac{1}{2}\mu r^2$, and the equation to be solved has the form:

$$(\hbar^2/2m)\left[\frac{1}{r}\frac{\partial}{\partial r}\left(r\frac{\partial \psi}{\partial r}\right) + \frac{1}{r^2}\frac{\partial^2 \psi}{\partial \phi^2}\right] + (E - \tfrac{1}{2}\mu r^2)\psi = 0. \quad (3)$$

The co-ordinates are separable, and it is convenient to write the solutions in the form

$$\psi = \frac{1}{r^{\frac{1}{2}}} f(r)\, e^{il\phi},$$

in which l is a positive or negative integer,† and to introduce a new measure of radial distance by writing ξ^2 for $2(m\mu)^{\frac{1}{2}} r^2/\hbar$. Then (3) becomes an equation for $f(\xi)$ alone:

$$f'' + [\varepsilon - (\tfrac{1}{2}\xi - \beta/\xi)^2] f = 0,$$

in which $\varepsilon = E/\hbar\omega_0 - \beta$ and $\beta^2 = l^2 - 1/4$. This is the same as (14.1), which need cause no surprise since the potential function in (14.1) arises in the classical treatment of radial motion of a conical pendulum, being the sum of the restoring potential $\frac{1}{2}\mu r^2$ and the centrifugal potential proportional to r^{-2}. The energy levels follow immediately from (14.5):

$$E = (2\nu + 1 + |l|)\hbar\omega_0, \tag{4}$$

with $\nu = 0$ or any positive integer. As expected from the solution in Cartesian co-ordinates, the levels are equally spaced with total zero-point energy $\hbar\omega_0$, $\frac{1}{2}\hbar\omega_0$ for each degree of freedom. All but the lowest level exhibit degeneracy, the level for which $E = n\hbar\omega_0$ being n-fold degenerate, since there are n ways in which $n-1$ units of $\hbar\omega_0$ may be distributed between the two Cartesian co-ordinates. It is this that allows different representations of the stationary-state wave-functions to be constructed by linear combination of the members of a degenerate set.

For the present argument the polar representation is suitable, providing a basis for synthesizing a wave-packet moving on a circular classical orbit. An excited state with $\nu = 0$ has the highest value of $|l|$ compatible with the energy, and therefore the highest value of β, so that the wave-function is kept as far from the origin as possible. The solution then follows the form (14.2) with $F(\xi)$ a constant:

$$\psi = C\, e^{-\frac{1}{4}\xi^2} \xi^{|l|}\, e^{il\phi}. \tag{5}$$

When $|l|$ is large, and the energy is $(|l|+1)\hbar\omega_0$, ψ is large only in the vicinity of $\xi_c = (2|l|)^{\frac{1}{2}}$, which is the classical radius for circular motion with energy $|l|\hbar\omega_0$, the energy of excitation above the ground state at $\hbar\omega_0$. Approximately, $\psi \sim e^{-\frac{1}{4}(\xi-\xi_c)^2}\, e^{il\phi}$, the Gaussian form of the radial variation being the same as in the ground state but centred on ξ_c. Already, therefore, we have proceeded halfway towards constructing the required wave-packet, having achieved a narrow Gaussian radial spread but with no angular restrictions; at this stage the wave-function may be likened to a model railway track arranged in a circle, with narrow radial wavefronts like the

† The axially symmetric case, $l = 0$, needs special treatment, but as it presents no difficulties and is not required for the present argument we shall ignore it.

sleepers. Angular confinement is effected by combining similar wave-functions having $\nu = 0$ and a range of values for l, so that at any instant constructive interference occurs only near some special value of ϕ. We may imagine the superposition of waves sweeping round very nearly the same circular track, but with a range of wavenumber, k, and of frequency, Ω. Since $k \propto l$ while $\Omega = E/\hbar$ and is linear in l, the group velocity $\mathrm{d}\Omega/\mathrm{d}k$ is independent of k. This is just the required condition for a wave-packet to propagate unchanged in form, which is what we set out to demonstrate.

Extension of the theory of a harmonic vibrator from one dimension to two has added virtually nothing that is not intuitively obvious, once it is understood how closely the classical and quantal treatments agree. Where classical arguments work in one dimension they work in two; where they fail, the extra dimension does nothing to redeem the failure.

Fermi resonance[1]: non-linear coupling[2]

It was noted in chapter 12 that vibrational energy levels in some molecules are significantly perturbed by parametric coupling, for example when a stretching mode has something like twice the frequency of a bending mode, as in CO_2. The effect is simply modelled as the motion of a particle in the potential well of fig. 2(*a*). If the valley were a parabolic minimum (elliptical contours, as in the preceding section) there would be independent normal modes of vibration parallel to the x and y axes; and if the frequency along y were exactly twice that along x the Lissajous figures for arbitrary initial conditions would be closed curves like those in fig. 2.17(*a*)–(*e*). Curvature of the valley, however, couples the x and y motions, and there are no longer any normal modes; they cannot be expected except with quadratic potentials, and now the potential contains cubic and quartic terms. For small curvature we may choose as our model potential

I.378

I.28

$$V = \tfrac{1}{2}\mu x^2 + 2\mu(y - cx^2)^2. \tag{6}$$

The valley, defined as the line on which $\partial V/\partial y = 0$, is the parabola $y = cx^2$. The classical equations of motion can be expressed in the form

$$\left.\begin{aligned} \ddot{\xi} + \xi - 8\xi(\eta - \xi^2) &= 0 \\ \text{and} \qquad \ddot{\eta} + 4(\eta - \xi^2) &= 0, \end{aligned}\right\} \tag{7}$$

in which $\surd m/\mu$ is taken as the unit of time, and $1/c$ as the unit of length.

The behaviour resulting from (7) is complicated, with exchange of energy between the ξ and η vibrations, and it is very helpful in appreciating the problem to make and study a real model. A simple pendulum in which the string is either replaced by, or incorporates, a spring capable of great extension is loaded until the extension of the spring is one-quarter the total length of the pendulum; then the vertical vibration has twice the frequency of the swing. No great care need be wasted on obtaining an exact ratio of

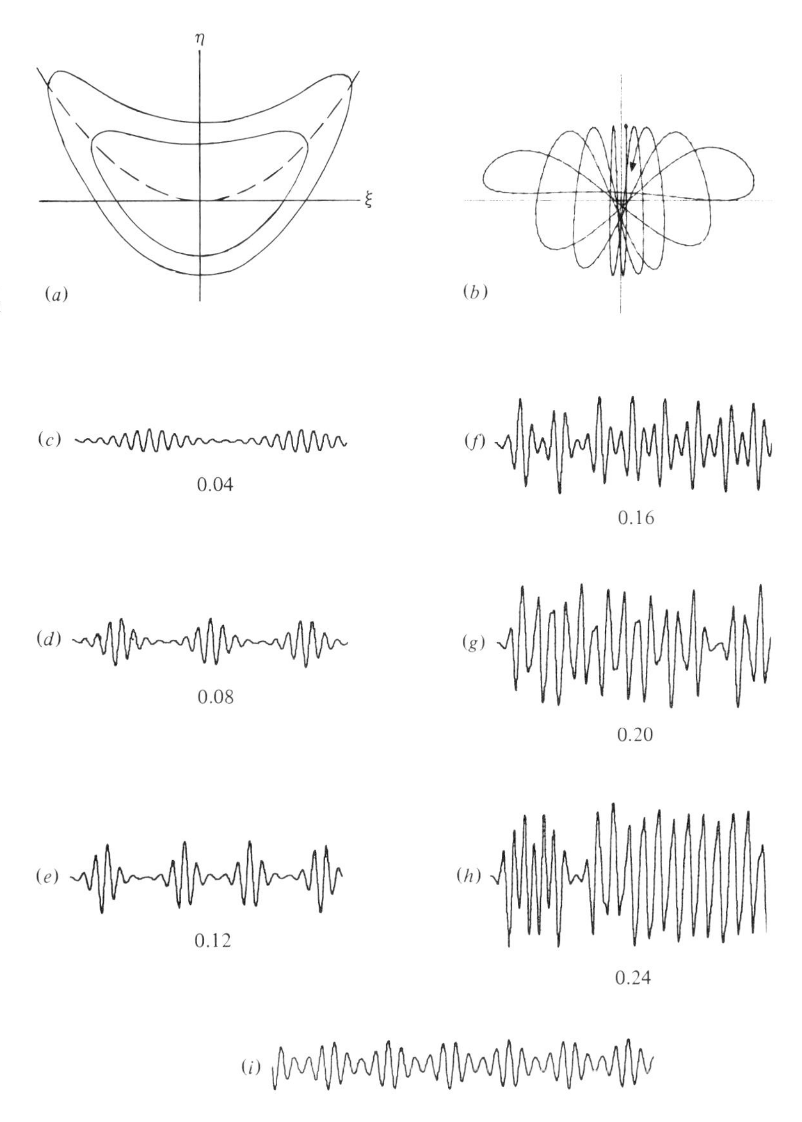

Fig. 2. (*a*) The anharmonic potential (6); the full lines are equipotentials and the broken line indicates the bottom of the valley. (*b*) Computed trajectory for the particle starting from rest at (0.01, 0.12). (*c*)–(*h*) $\xi(t)$ for the particle starting from rest at (0.01, η_0); values of η_0 are given below the curves. (*i*) $\xi(t)$ for the particle starting at the origin with initial velocity $(\dot{\xi}_0, \dot{\eta}_0) = (0.16, 0)$; the total energy is the same as in (*d*), but the beat period is noticeably shorter.

two between the frequencies – the behaviour is not affected by small departures from ideality. When the bob of the pendulum is pulled down and released, there is likely to be a component of the initial horizontal motion, albeit small, that is in the right phase to be amplified parametrically; amplification proceeds until the vertical motion has almost completely been converted into a horizontal swing. The variations of tension, at twice the frequency of swing, in their turn begin to re-establish the vertical vibration, and this resonant process proceeds until nearly all the sideways motion is lost again. The process then repeats itself almost in the same form; only the fact that each interchange of energy from one mode to the other is not quite complete gives rise to a slight variability from one 'beat' cycle to the next. Thus if it should happen that the sideways motion is reduced to a very small value, it will take correspondingly longer for the parametric process to build it up again. No such effect occurs with the resonant excitation of vertical motion by the variations of tension during a swing.

Fig. 2(b) is a trace of the initial process of transfer computed from (7). If the computation is allowed to proceed for many cycles all points within the contour $V=E$ are visited. This contrasts with the linear system, for which the energies of each normal mode are separately conserved, limiting the trajectory to points within a rectangle inscribed in the energy contour. The fact that there is only one energy invariant, not two, in the non-linear system does not imply, however, that all possible motions consistent with a given value of E are eventually to be found on the trajectory; on the contrary it appears that within the contour there is a continuous set of independent trajectories, passing through a given point with the same speed (as determined by E) but in different directions. That conclusion, at any rate, seems the natural interpretation of the traces in fig. 2(c) to (e), which are the ξ-component of motion after the system has been started from rest with various small displacements as described in the caption. At first sight one would think they were traces of the beat pattern characteristic of one component of a Lissajous figure, were it not that the beat frequency increases with amplitude. Trajectories having the same energy, but different initial conditions (e.g. fig. 2(d) and (i)), retain their distinct characters, just like Lissajous figures for different partitions of the same total energy between the two co-ordinates. Having stated this, however, one must enter a caveat; the higher energy trajectories, Fig. 2(f) to (h), show no stable pattern, even in a much longer sequence than is shown here. They may give the impression for a number of cycles of having settled down to regularity, but this proves illusory. It is possible that between (f) and (g) there is a critical value of the energy at which order gives way to chaos, but this question has not been investigated seriously. All the same, and however unprecisely, we shall use the term 'chaotic' for the higher energy trajectories.

I.270

Quantization of such a system may greatly reduce the scope for chaotic behaviour, especially when the mass is small and the levels correspondingly far apart. On the molecular scale, in fact, quite low excited levels already have energies that would lead to chaotic behaviour in a classical system. The quantal treatment is now markedly simpler than the classical and involves only elementary perturbation theory. It may be noted that the semi-classical approach offers no help here, since the motion is aperiodic and outside the scope of the Bohr–Wilson–Sommerfeld method, while the WKB solution runs into the same problems as the classical, that there is no explicit expression for the trajectories generated by (7).

32
37

The Hamiltonian derived from (6) has the form:

$$\mathscr{H}=(p_x^2+p_y^2)/2m+\tfrac{1}{2}\mu x^2+2\mu(y-cx^2)^2,$$

and leads to Schrödinger's equation:

$$\partial^2\psi/\partial X^2+\partial^2\psi/\partial Y^2+[\varepsilon-X^2-4(Y-\gamma X^2)^2]\psi=0, \tag{8}$$

in which the reduced co-ordinates are not the same as in (7). Here

$$\left.\begin{aligned}X=\alpha x=\xi/\gamma,\quad Y=\alpha y=\eta/\gamma,\quad \varepsilon=2E/\hbar\omega_0=(2E/\hbar)(m/\mu)^{\frac{1}{2}},\\ \gamma=c/\alpha\quad\text{and}\quad \alpha=(\mu/\hbar\omega_0)^{\frac{1}{2}}.\end{aligned}\right\} \tag{9}$$

In this form γ is seen as the coupling constant between the otherwise independent harmonic vibrations of X and Y. To relate the magnitudes in the classical and quantal treatments, note the suggestion that chaotic behaviour sets in when η_0, the initial value of η, is about 0.16. The energy of vibration is then $2\mu\eta_0^2/c^2$, so that $\varepsilon=4\mu\eta_0^2/\hbar\omega_0c^2=4\eta_0^2/\gamma^2\doteqdot 0.1/\gamma^2$. Thus if $\gamma=0.1$ the state $\varepsilon=10$, which is the second excited level of the Y-vibration at a frequency $2\omega_0$, has enough energy to cause trouble in the classical system. Let us bear this in mind when interpreting the outcome of the following first-order perturbation treatment of (8).

When $\gamma=0$, (8) is separable and ψ is the product of harmonic oscillator functions:

$$\psi(X,Y)=\psi_n(X)\psi_m(Y);\qquad \varepsilon=2n+4m+2. \tag{10}$$

The level $\varepsilon=2(N+1)$ has a degeneracy of $\frac{1}{2}N+1$ if N is even, and $\frac{1}{2}(N+1)$ if N is odd. The effect of a small non-vanishing γ is primarily to break this degeneracy, creating new states which are linear combinations of degenerate states. To first-order the admixture of other non-degenerate states may be neglected, and we may write as a trial solution

$$\psi_N(X,Y)=\lambda_0\psi_N(X)\psi_0(Y)+\lambda_1\psi_{N-2}(X)\psi_1(Y)+\cdots. \tag{11}$$

On substituting this in (8) we find

$$[\varepsilon-2(N+1)+8\gamma X^2Y](\lambda_0\psi_N\psi_0+\lambda_1\psi_{N-2}\psi_1+\cdots)=0, \tag{12}$$

if the term in γ^2 is dropped as irrelevant to the first-order calculation. Now multiply the left-hand side by $\psi_{N-2m}\psi_m$ and integrate over both variables:

$$\lambda_m\Delta+8\gamma\sum_l M^{(2)}_{N-2l,N-2m}M^{(1)}_{l,m}\lambda_l=0, \tag{13}$$

in which Δ is written for the displacement of the energy level, $\varepsilon-2(N+1)$, and $M^{(1)}$ and $M^{(2)}$ are the matrix elements as defined in (13.26) and (13.24). It should be noted, however, that $M^{(2)}$ relates here to an oscillator of frequency $2\omega_0$, and that correspondingly $M^{(2)}_{n,n-2}=\frac{1}{4}[n(n-1)]^{\frac{1}{2}}$. The only non-vanishing terms in (13) are those for which $l=m\pm 1$, and hence

$$\lambda_{m-1}\gamma a_m+\lambda_m\Delta+\lambda_{m+1}\gamma a_{m+1}=0, \tag{14}$$

where $a_m=[2m(N-2m+1)(N-2m+2)]^{\frac{1}{2}}$. This set of equations for the unknowns λ_m, with m running from 0 to $\frac{1}{2}N$ or $\frac{1}{2}(N-1)$, has a solution only if the determinant of the coefficients vanishes:

$$\begin{vmatrix} \Delta & \gamma a_1 & 0 & 0 & \cdots \\ \gamma a_1 & \Delta & \gamma a_2 & 0 & \cdots \\ 0 & \gamma a_2 & \Delta & \gamma a_3 & \cdots \\ 0 & 0 & \gamma a_3 & \Delta & \cdots \\ \vdots & \vdots & \vdots & \vdots & \end{vmatrix} = 0. \tag{15}$$

Hence the energy levels are determined: $\varepsilon = 2(N+1)+\Delta$. The splitting of the lowest levels is easily calculated. The lowest two, $N=0$ and 1, are non-degenerate and unperturbed to first order in γ. The next two, $N=2$ and 3, involve a 2×2 determinant and hence a quadratic equation for Δ; the levels split into doublets, while $N=4$ and 5 involve a cubic equation and split into triplets, and so on. The splitting is symmetrical and proceeds as in the following table:

N	ε (unperturbed)	Δ
0	2	0
1	4	0
2	6	$\pm 2\gamma$
3	8	$\pm 2\sqrt{3}\gamma$
4	10	0 and $\pm 4\sqrt{2}\gamma$
5	12	0 and $\pm 8\gamma$
6	14	$\pm 2.516\gamma$ and $\pm 10.662\gamma$
7	16	$\pm 4.059\gamma$ and $\pm 13.547\gamma$

Thenceforth the solution of (15) becomes progressively more tedious. One may guess, however, at the behaviour when N is large by assuming that the levels continue, as in the table, to be more or less evenly spaced; then the sum of the Δ^2 gives the spacing and overall spread. If N is even, multiplying out (15) yields an algebraic equation of the Nth order containing only even powers, and the sum of the Δ^2 is minus the coefficient of Δ^{N-2}, which is $\gamma^2\sum a_m^2$, or $\gamma^2 N^2(N+2)(N+4)/24$. For convenience let us take N to be a multiple of 4, so that evenly spaced levels would lie at $\Delta = 0$, $\pm\delta$, $\pm 2\delta$ etc. Then

$$\delta^2[1+4+9+\cdots+(N/4)^2] = \gamma^2 N^2(N+2)(N+4)/24.$$

Since the left-hand side sums to $\delta^2 N(N+2)(N+4)/192$, it follows that

$$\delta = (8N)^{\frac{1}{2}}\gamma. \tag{16}$$

Correspondingly the whole band of levels lies in thc range $2(N+1)\pm\gamma N^{\frac{3}{2}}/\sqrt{2}$.

The assumption of evenly spaced levels is consistent with the regular amplitude modulation exhibited by fig. 2(c)–(e). To make a quantitative comparison, we note that an initial displacement η_0 starts the system vibrating with ε equal to $4\eta_0/\gamma^2$, so that $N \sim 2\eta_0^2/\gamma^2$; hence from (16) the expected value of δ is $4\eta_0$ and the energy separation of the levels, i.e. $\frac{1}{2}\hbar\omega_0\delta$, is $2\hbar\omega_0\eta_0$. The resulting beat frequency should then be $2\omega_0\eta_0$, and the beat cycle should contain $1/2\eta_0$ cycles of the x-vibration. In the trace (*f*) in fig. 2 there are very nearly 3 cycles to the beat, entirely consistent with the initial value of 0.16 for η_0; and if the beats seem a little too long at the smaller values of η_0, the explanation probably lies in the very low amplitudes at the minima. Parametric amplification is a slow process in these circumstances but, as we have noted already in chapter 13, when the system is quantized the zero-point energy eliminates the early stages that are found in the classical system. The last trace, fig. 2(*i*), for which the energy is the same as in (*d*), shows that with a lesser degree of amplitude modulation the beat period is shorter; now, indeed, it is rather shorter than (16) predicts, $4\frac{1}{2}$ cycles rather than $6\frac{1}{4}$. The implication of this variation of beat period with depth of modulation is that the levels are not quite evenly spaced; and when a wave-packet is constructed to correspond to the classical solution different levels presumably are selected according to the initial conditions.

The conclusion of this comparison is that first-order perturbation theory works reasonably well at least until the beat frequency is about $\omega_0/3$, as in fig. 2(*f*), i.e. until the spacing of the levels is about one-third of the original separation of the degenerate levels. Clearly when N is large each band will have spread out to overlap many neighbouring bands, and it is perhaps surprising that they do not interact together so as to destroy all regularities. It may be noted, however, that even with the full perturbation expressed by $4(Y-\gamma X^2)^2$ in (8), the only components of the wave-function (11) that are coupled to components of $\psi_M(X, Y)$, belonging to another energy level, are those for which $|N-M|$ is 0, 2 or 4 and the m's differ by unity. The interaction between levels is consequently totally absent or not very strong, and the explanation for the modest success of the first-order calculation must surely lie here. The system deserves fuller investigation, both in its classical and its quantal aspects, than it has received so far.

Cyclotron orbits

A classical non-relativistic charged particle, moving freely in a uniform magnetic field $\boldsymbol{B}$, executes helical orbits at a frequency eB/m, independent of its energy and thus of the orbit radius $eB/(2mE)^{\frac{1}{2}}$. Its motion in the plane normal to $\boldsymbol{B}$, which is all we shall concern ourselves with, closely resembles that of a conical pendulum in a circular orbit and it is not surprising to find the quantum mechanics also very similar. Schrödinger's equation now contains the vector potential $\boldsymbol{A}$, defined by the equation curl $\boldsymbol{A}=\boldsymbol{B}$:

I.237

$$[(\hbar/i)\nabla - e\boldsymbol{A}]^2\psi/2m - E\psi = 0, \tag{17}$$

V being taken as zero everywhere. We shall return to the form of this equation later, but for the present let us take it as given. If $\boldsymbol{A}$ is chosen to

53

be non-divergent,

$$\hbar^2\nabla^2\psi/2m - \mathrm{i}e\hbar\boldsymbol{A}\cdot\nabla\psi/m + (E - e^2A^2/2m)\psi = 0. \tag{18}$$

The sensible choice of $\boldsymbol{A}$ depends on the co-ordinate system to be employed, and we shall consider two examples, $\boldsymbol{A} = (-By, 0, 0)$ appropriate to Cartesian co-ordinates and $\boldsymbol{A} = \frac{1}{2}\boldsymbol{B} \wedge \boldsymbol{r}$ appropriate to polar co-ordinates. Then (18) takes the alternative forms:

$$\hbar^2(\partial^2\psi/\partial x^2 + \partial^2\psi/\partial y^2)/2m + (\mathrm{i}e\hbar By/m)\partial\psi/\partial x + (E - e^2B^2y^2/2m)\psi = 0, \tag{19}$$

or

$$\hbar^2\left[\frac{1}{r}\frac{\partial}{\partial r}\left(r\frac{\partial\psi}{\partial r}\right) + \frac{1}{r^2}\frac{\partial^2\psi}{\partial\phi^2}\right]\Big/2m + (\mathrm{i}e\hbar B/m)\partial\psi/\partial\phi + (E - e^2B^2r^2/8m)\psi = 0 \tag{20}$$

In both cases the variables are separable. In (19) we substitute $\psi = Y(y)\,\mathrm{e}^{\mathrm{i}kx}$ and find that Y must obey the equation:

$$\hbar^2 Y''/2m + [E - e^2B^2(y + \hbar k/eB)^2/2m]Y = 0, \tag{21}$$

which is the harmonic oscillator equation with the natural frequency $\omega_\mathrm{c} = eB/m$ and with the attractive centre at the point $y = -\hbar k/eB$. For a large rectangular slab k can take many values to allow $\mathrm{e}^{\mathrm{i}kx}$ to satisfy periodic boundary conditions in the x-direction, while still allowing the oscillator function to fit within the bounds of the sample in the y-direction. Each energy level $(n+\frac{1}{2})\hbar\omega_\mathrm{c}$ is thus highly degenerate, so that infinitely many alternative representations of the wave-functions are possible. It is not to be expected that any one choice, determined by the mathematical technique of solving the equation, will bear a close resemblance to the classical orbit pattern. In this particular case the y-variation reproduces the characteristic pattern of harmonic motion, as if there had been superimposed a lot of wave-functions closely confined to the classical orbit, but centred at different points along the line $y = -\hbar k/eB$, with their phases adjusted to give a progressive phase variation $\mathrm{e}^{\mathrm{i}kx}$.

Wave-functions resembling a classical orbit are not merely hypothetical; one such arises from the solution of (20), from which the ϕ-variable may be separated by the substitution $\psi = (1/r^{\frac{1}{2}})f(r)\,\mathrm{e}^{\mathrm{i}l\phi}$. Once more (14.1) emerges, with $\xi = |eB/\hbar|^{\frac{1}{2}}r$, $\varepsilon = 2E/\hbar\omega_\mathrm{c} \pm l - \beta$ and $\beta^2 = l^2 - \frac{1}{4}$. Hence **45**

$$E = (2\nu + 1 + |l| \mp l)\tfrac{1}{2}\hbar\omega_\mathrm{c}, \tag{22}$$

the sign being that of the charge on the particle. The same level structure emerges, of course, but the curious appearance of $|l| \mp l$ reflects the basic difference between cyclotron motion and a conical pendulum, in that only one sense of rotation is possible in the former. When $\nu = 0$, which gives a wave-function most nearly confined to the classical orbit, only positive values of l are permitted for positively charged particles, i.e. only anticlockwise circulation as viewed along the direction of $\boldsymbol{B}$, and only negative l, clockwise circulation, for negative particles. Otherwise the solution is like that for the conical pendulum, with the particle closely confined to the classical orbit when $|l|$ is large. Moreover, like the pendulum, stable wave-packets may be constructed to localize the particle on its orbit, and keep it so localized. This should give encouragement to those who are accustomed to think of the motion of particles in classical terms; when dealing with the dynamical behaviour of conduction electrons in metals and semiconductors, for example, the classical mode of thought is virtually a necessity on account of the complexities introduced by the crystal lattice. Quantum mechanics serves to show how the electron can travel through a perfect lattice without suffering collisions with individual atoms, and it can produce expressions for the modifications to the dynamical properties caused by the lattice. Once it has shown the character of the equivalent classical particle which would simulate these dynamical properties there is some merit in forgetting quantum mechanics and thinking of the classical trajectories of these particles as they move in electric and magnetic fields, and suffer scattering from defects of the lattice. The fact that one can construct stable wave-packets that follow the classical trajectory helps one to believe that the undoubted success of classical thinking applied to these problems may not be merely a lucky accident.

This is not to suggest of course, that it is always safe to neglect the discrete energy level structure when a magnetic field is present, but this is something that can often be supplied as an afterthought through the semi-classical process of phase-integral quantization, already shown to be a great value for anharmonic oscillators. If a free particle in its cyclotron orbit is allowed to present close analogies with a harmonic vibrator, the modified dynamics resulting from the crystal lattice may be seen as analogous to anharmonicity, often of extraordinary complexity; but this proves to present no obstacle to semi-classical quantization. A number of interesting and important points arising in the pursuit of this line of thought are all traceable to a rather strange effect that is already to be found in the behaviour of a free particle. The nearly-classical orbit, in which $\nu = 0$ and l is large has a perimeter of $2\pi v/\omega_c$, if v is the velocity of a classical particle of the same energy, $l\hbar\omega_c$. The wave that sweeps round it has l wavelengths fitting into the circuit. Hence the wavelength, measured round the orbit, is $2\pi v/\omega_c l$, i.e. $2h/mv$, since $l = \frac{1}{2}mv^2/\hbar\omega_c$. This wavelength is *twice* the value that one would naively expect for a de Broglie wave.

32

Quantization in a magnetic field

The explanation lies in the classical mechanics of a charged particle in a magnetic field. The non-dissipative Lorentz force can be incorporated in the Lagrangian formalism by redefining the momentum conjugate to the position variable; instead of being merely $m\boldsymbol{v}$, $\boldsymbol{p}$ must be taken as $m\boldsymbol{v} + e\mathbf{A}$, so that the Hamiltonian form of the energy is $(\boldsymbol{p} - e\mathbf{A}) \cdot (\boldsymbol{p} - e\mathbf{A})/2m + V(\boldsymbol{r})$. The transition to Schrödinger's equation, by the substitution $\boldsymbol{p} = \hbar\nabla/\mathrm{i}$, leads to (17). In other words, the de Broglie wavelength is determined by $m\boldsymbol{v}+e\boldsymbol{A}$ rather than by $m\boldsymbol{v}$ alone, and in the case just discussed $e\boldsymbol{A}$ is antiparallel to $m\boldsymbol{v}$ and half its magnitude; for a positively charged particle moving anticlockwise in an orbit of radius r, $mv = eBr$, while the vector potential runs clockwise and has magnitude $\frac{1}{2}Br$.

Since the addition to $\boldsymbol{A}$ of $\nabla\beta$, β being any differentiable scalar function of $\boldsymbol{r}$, leaves curl $\boldsymbol{A}$ and therefore $\boldsymbol{B}$ unaltered, the local de Broglie wavelength can be modified at will, but this has no effect on the quantization of the orbits since the total phase change around the orbit is not changed. The two choices of gauge for $\boldsymbol{A}$ already discussed in connection with (19) and (20) illustrate this point. A detailed appreciation of the motion of electrons in a metal under the influence of $\boldsymbol{B}$ also depends on recognizing the contribution of $\boldsymbol{A}$ to the phase. If the periodic field of the lattice is very weak electrons moving in almost any direction hardly notice its existence, but it matters considerably if it can cause Bragg reflection; for this to occur there must be a Fourier component in the periodic lattice field having wavenumber $\boldsymbol{g}$ such that an electron with wavenumber $\boldsymbol{k}$ can be reflected into a state of the same energy and of wavenumber $\boldsymbol{k}+\boldsymbol{g}$. Very commonly, as in X-ray diffraction, the satisfying of the Bragg condition is achieved by making the magnitudes of $\boldsymbol{k}$ and $\boldsymbol{k}+\boldsymbol{g}$ the same, but this is not an essential requirement, and we have a particular case here where the more general statement of Bragg's law is required. An electron in its orbit may find itself reflected by a set of lattice planes into another that is differently centred (fig. 3), and under the right conditions one or more

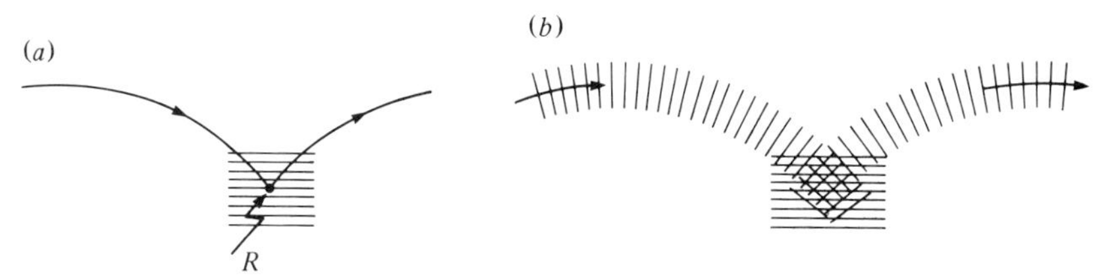

Fig. 3. (*a*) An electron moving in a uniform magnetic field suffers Bragg reflection at R from the lattice planes, shown as horizontal lines. (*b*) Schematic, but incorrect (see text), representation of Bragg reflection in a magnetic field.

further reflections may bring it back to the original orbit; an example will be found in the quasi-classical diagram, fig. 6. In line with our discussion of localized orbit wave-functions we might imagine the corresponding quantum-mechanical picture to take a form like that in fig. 3(*b*), and the process of Bragg reflection to act on a wavefront of limited extent, but still much wider than the lattice spacing. There seems little objection to this view except that it is incorrect to represent both sets of wavefronts as radial. For this is to imply that each orbit is gauged with its own centre as gauge centre,† while in practice we ought to choose the same gauge centre for both. This leads us to derive the rule for phase variation around an orbit which is gauged with respect to a point not its centre.

† If $\boldsymbol{A}$ is written as $\frac{1}{2}\boldsymbol{B}\wedge(\boldsymbol{r}-\boldsymbol{R})$, $\boldsymbol{R}$ is the position of the gauge centre round which $\boldsymbol{A}$ circulates.

If ψ_0 is a solution of (17) for a given value of E, the addition of $\nabla\beta$ to $\boldsymbol{A}$ can be compensated by replacing ψ_0 by $\psi_o\,\mathrm{e}^{\mathrm{i}e\beta/\hbar}$; it is readily verified that the latter is now a solution for the same E. If then we express ψ_0 as $f(\boldsymbol{r})\,\mathrm{e}^{\mathrm{i}\theta(\boldsymbol{r})}$, $f(\boldsymbol{r})$ being real, the change of gauge alters the phase factor without changing the amplitude, the new solution being $f(\boldsymbol{r})\,\mathrm{e}^{\mathrm{i}(\theta+e\beta/\hbar)}$. This result can be expressed otherwise in terms of the local wavenumber $\boldsymbol{k}$, defined as the gradient of phase. Originally $\boldsymbol{k}_0=\nabla\theta$, and the change of gauge converts $\boldsymbol{k}_0$ to $\boldsymbol{k}_0+e\nabla\beta/\hbar$, or $\boldsymbol{k}_0+e\Delta\boldsymbol{A}/\hbar$, where $\Delta\boldsymbol{A}$ is the change in $\boldsymbol{A}$ at the point considered. The significance of this rule is immediately apparent when applied to a wave-function which locally has the form of a plane wave, $\mathrm{e}^{\mathrm{i}\boldsymbol{k}_0\cdot\boldsymbol{r}}$, and which describes a particle of momentum $\hbar\boldsymbol{k}_0$. Now this momentum is not $m\boldsymbol{v}$ but the conjugate of the position vector, i.e. $\hbar\boldsymbol{k}_0=m\boldsymbol{v}+e\boldsymbol{A}$. The simultaneous addition of $\Delta\boldsymbol{A}$ to $\boldsymbol{A}$ and $e\Delta\boldsymbol{A}/\hbar$ to $\boldsymbol{k}_0$ leaves $m\boldsymbol{v}$ unchanged, so that the same dynamical behaviour is successfully described in the new representation.

For the particular gauge change in which the gauge centre is shifted from the origin to $\boldsymbol{R}$, $\Delta\boldsymbol{A}=\frac{1}{2}\boldsymbol{B}\wedge(\boldsymbol{r}-\boldsymbol{R})-\frac{1}{2}\boldsymbol{B}\wedge\boldsymbol{r}=-\frac{1}{2}\boldsymbol{B}\wedge\boldsymbol{R}$, which is constant. Thus every $\boldsymbol{k}_0$ is increased by the same vector increment, $-\frac{1}{2}e\boldsymbol{B}\wedge\boldsymbol{R}/\hbar$. For an orbit whose centre is the origin, $\boldsymbol{k}_0$ at a point $\boldsymbol{r}$ is originally $e\boldsymbol{r}\wedge\boldsymbol{B}/2\hbar$, and if the gauge centre is moved to $\boldsymbol{r}$ itself the increment to $\boldsymbol{k}_0$ is equal to $\boldsymbol{k}_0$ itself, so that now $\boldsymbol{k}_0=e\boldsymbol{r}\wedge\boldsymbol{B}/\hbar$. This is just $m\boldsymbol{v}/\hbar$, as if there were no magnetic field present, but it applies only to that point on the orbit which is the new gauge centre. At the opposite end of the diameter $\boldsymbol{k}$ is reduced to zero and, of course, $\oint\boldsymbol{k}\cdot\mathrm{d}\boldsymbol{r}$ round the orbit is invariant with respect to gauge changes, since $\oint\Delta\boldsymbol{A}\cdot\mathrm{d}\boldsymbol{r}=0$. It is now clear that the condition for Bragg reflection is essentially unchanged by the presence of $\boldsymbol{B}$. For if we choose the point of reflection, R in fig. 3(a), as the gauge centre, the local wavenumber on each orbit is determined by $m\boldsymbol{v}$ as if no field were present. Reflection occurs when the electron in its orbit happens to move in such a direction as would cause it to suffer reflection when $\boldsymbol{B}=0$. This result is naturally independent of the choice of gauge centre; since shifting the centre by $\boldsymbol{R}'$ adds the same vector increment to the wave-vectors on the two orbits, their difference remains equal to $\boldsymbol{g}$ whatever may happen to them individually.

Quantization of non-circular orbits in real metals

It is possible to proceed from this point to calculate the phase length round an orbit composed of a number of circular arcs joined by Bragg reflections, and hence to derive the quantization rule, which turns out to be more conveniently expressed in terms of the area enclosed by the orbit than in terms of its perimeter. The argument, which demands rather careful exposition, will not be given here.[3,4] It is sufficient to remark that it agrees with an earlier treatment due to Onsager,[5] of great elegance and economy, which by-passes the more awkward problems by going back to the semi-classical method of phase-integral quantization. We have seen how well this works with highly anharmonic oscillators, and equal success attends its application to the cyclotron orbits of electrons, even when they move in lattice fields strong enough for the assumption of nearly free motion between Bragg reflections to be quite inadequate. Onsager assumed that the electron could be treated as a classical particle executing a well-defined orbit, but with modified dynamical properties. This has been to a certain degree, though far from rigorously, justified by later analyses.[6] In the absence of a magnetic field the energy is taken to be a function, possibly very complicated, of momentum $\boldsymbol{p}$; $E = E(\boldsymbol{p})$ generalizes the Newtonian form $E = p^2/2m$. Surfaces of constant E plotted in a three-dimensional $\boldsymbol{p}$-space serve to represent the form of this function, and in real metals these energy surfaces are normally very different indeed from the concentric spheres that describe a Newtonian particle. A particular example is shown in fig. 4 to illustrate the need for an approach to these problems that is

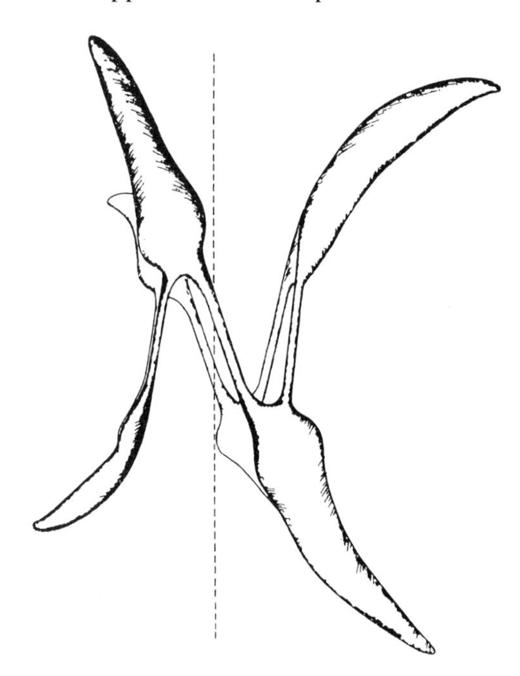

Fig. 4. Fermi surface of arsenic, as derived by Lin and Falicov[7].

not too strongly dependent on simplified analytical representations. The analogue to the de Broglie wave associated with a particle of momentum $\boldsymbol{p}$, and wavenumber $\boldsymbol{k} = \boldsymbol{p}/\hbar$, is a Bloch wave to which also wavenumber $\boldsymbol{k}$ may be assigned. Thus the surfaces representing $E(\boldsymbol{p})$ serve equally for $E(\boldsymbol{k})$ if the scale is altered. Since in fact the meaning of $\boldsymbol{k}$ for a Bloch wave is rather clearer than the meaning of $\boldsymbol{p}$, it is customary to proceed in terms of $\boldsymbol{k}$ rather than $\boldsymbol{p}$; but one must remember that although, when a magnetic field is applied, the same function $E(\boldsymbol{k})$ still defines the energy, $\boldsymbol{k}$ no longer defines the local wavenumber, which must now be taken as $\boldsymbol{k} + e\mathbf{A}/\hbar$. This does not, however, affect the group velocity which is still that of a wave of frequency $\omega = E(\boldsymbol{k})/\hbar$, and serves to determine the velocity $\boldsymbol{v}$ of a particle in the state $\boldsymbol{k}$. For one-dimensional motion the group velocity is $\mathrm{d}\omega/\mathrm{d}k$, i.e. $(\mathrm{d}E/\mathrm{d}k)/\hbar$, and the same applies to any component of the three-dimensional velocity; $\boldsymbol{v}$ therefore has the Cartesian form $\hbar^{-1}(\partial E/\partial k_x, \partial E/\partial k_y, \partial E/\partial k_z)$, i.e. $\boldsymbol{v} = \hbar^{-1}\nabla_k E$, where ∇_k is the vector derivative (gradient) with respect to position in $\boldsymbol{k}$-space. There is no component of $\nabla_k E$ lying in the constant energy surfaces, and $\boldsymbol{v}$ is therefore directed normal to these surfaces with a magnitude varying inversely as the separation of neighbouring surfaces, E and $E + \delta E$.

In a uniform magnetic field $\boldsymbol{B}$ the classical equation of motion for an electron under the influence of the Lorentz force takes the form $\dot{\boldsymbol{p}} = e\boldsymbol{v} \wedge \boldsymbol{B}$, and this result still holds, so that $\boldsymbol{k}$ changes according to the equation:

$$\dot{\boldsymbol{k}} = (e/\hbar)\boldsymbol{v} \wedge \boldsymbol{B}. \tag{23}$$

All changes in $\boldsymbol{k}$ are in the plane normal to $\boldsymbol{B}$ and, being also normal to $\boldsymbol{v}$, cause the point representing $\boldsymbol{k}$ for the electron to move round a constant energy surface. If we cut a plane section of the surface normal to $\boldsymbol{B}$ the resulting curve is the $\boldsymbol{k}$-trajectory of the electron. In general there will be a component of $\boldsymbol{v}$ parallel to $\boldsymbol{B}$, but we shall ignore this longitudinal motion and concentrate on the transverse component $\boldsymbol{v}_\mathrm{t}$, normal to $\boldsymbol{B}$. Let $\boldsymbol{v}_\mathrm{t}$ in (23) be written as $\dot{\boldsymbol{r}}_t$; then $\dot{\boldsymbol{r}}_t$ is related to $\dot{\boldsymbol{k}}$ by the process of multiplying by $\hbar/eB$ and turning through $\pi/2$. So too, on integrating, $\boldsymbol{r}_t$ is related to $\boldsymbol{k}$ except for an arbitrary integrating constant. As illustrated in fig. 5, when

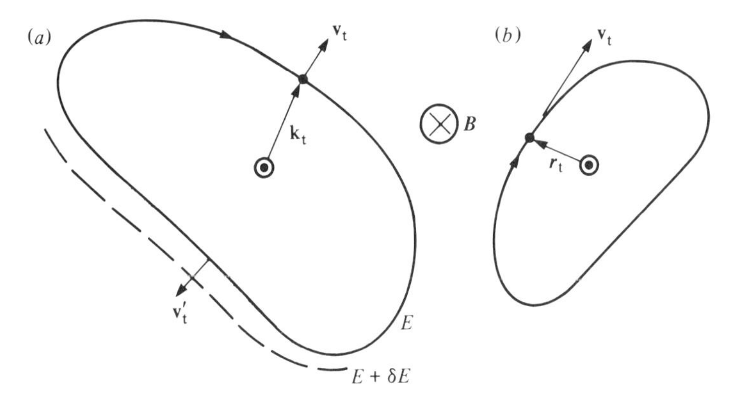

Fig. 5. Motion of an electron in a uniform magnetic field: (*a*) in $\boldsymbol{k}$-space (*b*) in real space. The curves are similar in shape, but oriented at 90° to one another.

the section of the constant energy surface is a closed curve, the electron whose $\boldsymbol{k}$ moves round this will describe in real space a similar closed orbit, which may be located anywhere and whose size is inversely proportional to the magnetic field strength. Strictly, when we take motion along $\boldsymbol{B}$ into account, this orbit is the projection on a plane normal to $\boldsymbol{B}$ of an irregular helical trajectory which repeats itself exactly every time round.

The cyclotron frequency ω_c of an orbit may be expressed in geometrical terms by calculating how long it takes to sweep out the annular area between the two neighbouring sections, E and $E + \delta E$, in fig. 5. The representative point and its vector $\boldsymbol{v}_t'$ sweep around at a rate $\dot{k} = eBv_t'/\hbar$, from (23); the width of the annulus is $\delta E/\hbar v_t'$ and the annular area is therefore swept out at the constant rate $eB\delta E/\hbar^2$. If the area of the section of the constant energy surface is $\mathscr{A}_k(E)$, the annulus has area $\mathrm{d}\mathscr{A}_k/\mathrm{d}E \cdot \delta E$ and is swept out in a time $(\hbar^2/eB)\,\mathrm{d}\mathscr{A}_k/\mathrm{d}E$.

Hence $$\omega_c = (2\pi eB/\hbar^2)/(\mathrm{d}\mathscr{A}_k/\mathrm{d}E). \qquad (24)$$

For the special case of free electrons moving in the plane normal to $\boldsymbol{B}$, $E = \hbar^2k^2/2m$, so that $\mathscr{A}_k = 2\pi mE/\hbar^2$ and (24) gives eB/m for ω_c, as expected. If the quantized levels are separated in energy by $\hbar\omega_c$, they are represented on the $\boldsymbol{k}_t$-plane by a set of nesting curves, like those in fig. 5, each the section of a constant energy surface. The areas enclosed by the curves increase in equal steps of $2\pi eB/\hbar$, independent of any parameters relating to the particular metal. This is reminiscent of the phase-integral for vibrators, and is indeed confirmed by the following direct evaluation of the phase-integral. **34**

Let us choose corresponding points in fig. 5 as origins for $\boldsymbol{k}_t$ and $\boldsymbol{r}_t$. Then when the electron is at a point $\boldsymbol{r}_t$, $\boldsymbol{k}_t$ has a value $e\boldsymbol{r}_t \wedge \boldsymbol{B}/\hbar$. This must be supplemented by $e\boldsymbol{A}/\hbar$ to give $\boldsymbol{k}_t'$ determining the phase variations. Consequently the phase length of the orbit is expressed in the form

$$J = \hbar \oint \boldsymbol{k}_t' \cdot \mathrm{d}\boldsymbol{r}_t = e \oint (\boldsymbol{r}_t \wedge \boldsymbol{B} + \boldsymbol{A}) \cdot \mathrm{d}\boldsymbol{r}_t = e\left\{ -\boldsymbol{B} \cdot \oint \boldsymbol{r}_t \wedge \mathrm{d}\boldsymbol{r}_t + \Phi \right\}, \qquad (25)$$

where $\Phi = \oint \boldsymbol{A} \cdot \mathrm{d}\boldsymbol{r}_t$ and is the flux of $\boldsymbol{B}$ passing through the orbit; if $\mathscr{A}_r$ is the area of the orbit in real space, $\Phi = B\mathscr{A}_r$. For the first term in (25) we note that $\oint \boldsymbol{r}_t \wedge \mathrm{d}\boldsymbol{r}_t$ is just $2\mathscr{A}_r$, so that the first term is twice the second, and

$$J = -eB\mathscr{A}_r, \qquad (26)$$

application of the quantization rule (14.12) then yields the allowed values of $\mathscr{A}_r$:

$$\mathscr{A}_r = (n + \gamma)2\pi\hbar/eB. \qquad (27)$$

The allowed areas in real space increase in equal steps of $2\pi\hbar/eB$, and correspondingly the allowed values of $\mathscr{A}_k$ increase in steps $(eB/\hbar)^2$ times larger, i.e. $2\pi eB/\hbar$, to give, according to (24), the expected energy intervals of $\hbar\omega_c$. The allowed orbits in real space are those whose flux content is $(n + \gamma)2\pi\hbar/e$. The unit of flux, $2\pi\hbar/e$, is sometimes referred to as the flux quantum, and in many respects it behaves as its name suggests, in that when it is possible to isolate a bundle of flux lines the quantum conditions applying to the system that effects this isolation are rather likely to impose a value on Φ that is an integral multiple of $2\pi\hbar/e$. The easiest way of achieving this isolation of Φ, however, is by enclosure within a superconducting ring, but in this case it is a multiple of the half-quantum $\pi\hbar/e$ that may be trapped. The concept of a flux quantum therefore needs to be treated with caution – it is not so absolute an entity as most quanta.

Onsager's demonstration of the quantization of cyclotron orbits in terms of their area alone, whatever the details of their shape, has had profound consequences for the study of metals. The de Haas–van Alphen effect,[8] whose essential feature is the oscillatory variation with magnetic field strength of the magnetization of a metal single crystal at low temperatures, is a direct consequence on the orbit quantization. By measuring the period of the oscillation the areas of certain dominant sections of the Fermi surface (the constant energy surface marking the division between filled and unfilled states) may be determined with great accuracy, and as the direction of $\boldsymbol{B}$ is changed other sections come under study. Enough information has been amassed for many metals[9] to allow the detailed shape of their Fermi surfaces to be constructed; fig. 4 is an example. This is the first step towards acquiring a full specification of the dynamical properties of the conduction electrons, and the systematic pursuit of this goal has been rewarded by a great advance in knowledge of metallic behaviour. Further, it has brought about greatly improved calculations from first principles of the shape to be expected for the Fermi surface in a given metal. Solution of Schrödinger's equation for electrons moving in the potential of the ionic lattice has proved too complicated to accomplish without making approximations, or introducing simplified models in place of the real ionic potentials. Knowledge of the required answer has provided an essential test of the theoretical simplifications, as a result of which these approximate methods may be applied confidently to more advanced problems where direct experimental confirmation is harder to acquire.

Magnetic breakdown[10]

The periodic field of the ionic lattice, through which the conduction electrons move, does not merely modify their orbits in matters of detail; it produces significant changes in the nature of the orbits, such as reversing their sense. The hexagonal two-dimensional pattern of fig. 6 illustrates this point. If the lattice potential is negligible, any of the circular orbits shown is one out of a myriad of possible free-electron orbits, with their centres arbitrarily located. As the potential is increased in strength, however, it is convenient to select for examination a subset, as in the diagram, chosen so that at all the intersections and only there, the Bragg condition is satisfied; the electrons cannot now complete their circular orbits but instead, depending on where they started, they may either traverse small triangular orbits ($\mathscr{T}$) in the same sense, or larger hexagonal orbits ($\mathscr{H}$) in a retrograde sense. The Onsager quantization rule (27) determines at which electron energies orbits of each kind are allowed. It may be noted that for a given magnetic field strength the free-electron orbits which are coupled by Bragg reflection are defined by the same lattice of orbit centres for all energies;† as E

† The lattice of orbit centres is the same as the reciprocal lattice, but scaled by $\hbar/eB$.

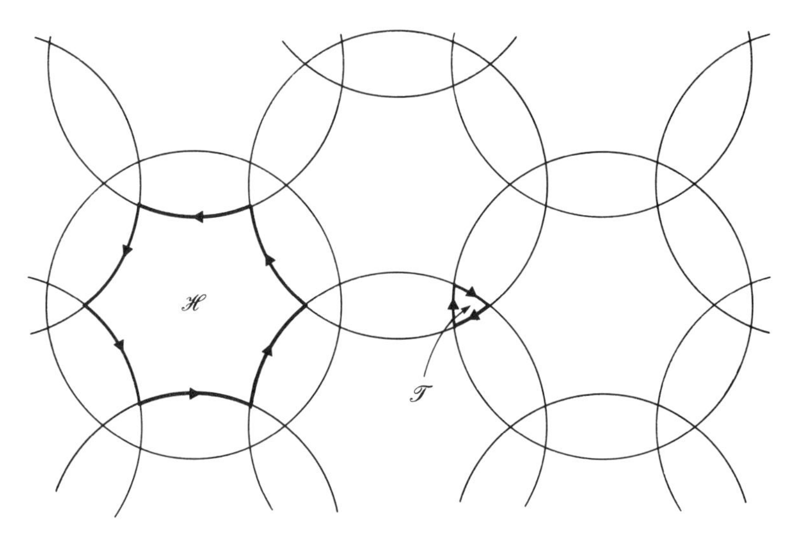

Fig. 6. Free electron orbits coupled by Bragg reflection from a hexagonal lattice, showing the network of paths made available by magnetic breakdown.

increases, therefore, and the circles get larger, the triangular orbits expand while the hexagonal orbits contract. The permitted energy levels then fall into two sets, not evenly spaced like free-electron levels (since the areas are not linear in energy) and as B is increased the levels for $\mathscr{T}$ move upwards in energy while those for $\mathscr{H}$ move downwards.

After this brief preliminary, let us ask how the transition from the free-electron orbits to $\mathscr{T}$ and $\mathscr{H}$ is effected as the lattice potential is gradually strengthened. The answer is simple in principle – with a weak lattice potential Bragg reflection is incomplete, and at every junction an electron has a certain probability of reflection, or alternatively may carry on round the free-electron orbit to the next junction, where the same choice arises. The process of Bragg reflection may be pictured as the production of a weak reflected wavelet by each lattice plane as the incident wave passes. At the correct angle of incidence all reflected wavelets add in phase to give a strong resultant – indeed if the incident wave is not deflected by a magnetic field reflection is total for angles of incidence within a certain narrow range round the Bragg angle. But the magnetic field, by bending the wave path, limits the length within which the Bragg condition is nearly enough satisfied, and if there is not enough reflected amplitude from the lattice planes within this length reflection will be only partial. The stronger the field the more likely the electron is to get through without reflection. This is the phenomenon of *magnetic breakdown.*

Under ideal conditions of purity and crystalline perfection the electrons may travel considerable distances without loss of phase information, and it is now essential to take account of interference between waves that have traversed different paths. The whole network in fig. 6 may be thought of as a system that can guide waves in the direction of the arrows and which will show a multitude of resonances, since each possible closed path has something of the character of a circulating resonator. Suppose, for example, that Bragg reflection is nearly complete; then a wave circulating anticlockwise on a hexagonal orbit will excite a small amplitude in each neighbouring triangular orbit, and the wave fed into the triangle will run round inside it, with only slight leakage into its contiguous hexagons. If the phase length of the triangle is an exact multiple of 2π it will resonate – the weak amplitude fed in from the hexagon will build up to such a resultant that ultimately there will be a significant extraction from the hexagon and transfer of wave amplitude into the other orbits in the vicinity. In this way the electron originally described as confined to a single hexagonal orbit can find its way to others and continue in a sort of diffusive motion to wander through the metal. As the effect depends on the electron's energy matching one of the permitted levels for the triangular orbits (this is the meaning of resonance in this context), the paths available to an electron may at some magnetic field strengths be closely confined to an orbit such as $\mathscr{H}$, and at others allow considerable excursions to distant regions. As a result, the conductivity of the metal, which is determined by how far an electron may travel, and in which directions, shows rapid oscillations with strength of magnetic field, as illustrated in fig. 7. The phenomenon of

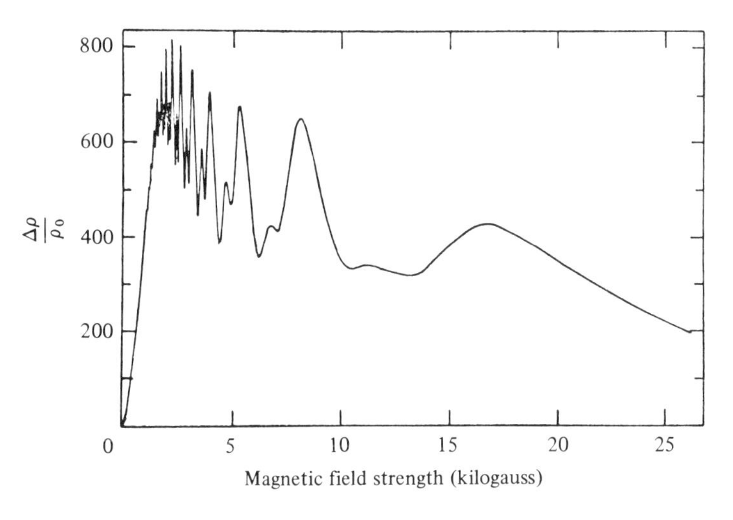

Fig. 7. Magnetoresistance in a pure zinc single crystal at 4 K (Stark[11]). Current flow is in the plane normal to the hexagonal axis, which is the direction of the applied magnetic field, $\boldsymbol{B}$; $\Delta\rho/\rho_0$ is the change in resistance of the sample expressed in terms of the resistance when $\boldsymbol{B} = 0$.

magnetic breakdown is not uncommon in metals; that is to say, there is a considerable number of examples where Bragg reflection is weak enough to allow breakdown in experimentally attainable magnetic fields. The details of what happens then depend critically on the arrangements of coupled orbits, and very few general rules have emerged from analyses of special cases.

There is one general result of some interest, however, and not solely in the context of magnetic breakdown, since it depends on a viewpoint that can be applied in other cases. We have seen that when the cyclotron orbits are well-defined the permitted energy levels form a discrete ladder. When, however, they are coupled by magnetic breakdown the levels spread into something more like a continuum, though it may still show gaps. A theorem due to Falicov and Stachowiak[12] provides a clue as to the character of the energy level diagram. Let us, at $t=0$, establish a wave-packet at the point $\boldsymbol{r}_0$ in the metal, and let this wave-packet be ideally small: $\Psi(r_0, 0) = \delta(\boldsymbol{r}-\boldsymbol{r}_0)$. It will immediately begin to spread, and the manner of its development may be written down formally by expressing $\delta(\boldsymbol{r}-\boldsymbol{r}_0)$ as a Fourier sum over all stationary states of the electron in the magnetic field:

$$\delta(\boldsymbol{r}-\boldsymbol{r}_0)=\sum_n \psi_n^*(\boldsymbol{r}_0)\psi_n(\boldsymbol{r}), \tag{28}$$

the amplitude of the nth wave-function being simply $\psi_n^*(\boldsymbol{r}_0)$. At time t, therefore, the development of each wave-function gives rise to the resultant

$$\Psi(\boldsymbol{r}, t)=\sum_n \psi_n^*(\boldsymbol{r}_0)\psi_n(\boldsymbol{r})\,\mathrm{e}^{-\mathrm{i}E_nt/\hbar}. \tag{29}$$

In particular, at the point $\boldsymbol{r}_0$ where the wave-packet started,

$$\Psi(\boldsymbol{r}_0, t)=\sum_n \psi_n^*(\boldsymbol{r}_0)\psi_n(\boldsymbol{r}_0)\,\mathrm{e}^{-\mathrm{i}E_nt/\hbar}, \tag{30}$$

and if we take the average value of $\Psi(\boldsymbol{r}_0, t)$ over all initial starting points, $\boldsymbol{r}_0$, the normalization of the ψ_n ensures that each contributes the same average value of $\psi_n^*\psi_n$. Hence,

$$\overline{\Psi(\boldsymbol{r}_0, t)}\propto\sum_n \mathrm{e}^{-\mathrm{i}E_nt/\hbar}. \tag{31}$$

Now the density of states $g(E)$ is the number of states per unit energy range at E, and its Fourier transform G may be defined as

$$G(x)=\int g(E)\,\mathrm{e}^{-\mathrm{i}Ex}\,\mathrm{d}E=\sum_n \mathrm{e}^{-\mathrm{i}E_nx}. \tag{32}$$

Comparison of (31) and (32) shows that the time-development of $\overline{\Psi(\boldsymbol{r}_0, t)}$, when Fourier-analysed, yields the density of states. There are several subtle points of difficulty which have been glossed over in this brief résumé, but enough has been said to allow a qualitative discussion, and the original paper may be consulted for details.

Let us use the free electron in two dimensions to illustrate this result. The initial wave-packet spreads out in all directions as if a cluster of classical electrons had been released at $\boldsymbol{r}_0$. No matter what their original direction, they will all return to $\boldsymbol{r}_0$ after one cyclotron period, and we expect the wave-function to show a sharp peak as it reconstitutes itself, only to spread once more, but at subsequent equal intervals to return. Thus $\overline{\Psi(\boldsymbol{r}_0, t)}$ is expected to present an even sequence of pulses with a period of $2\pi/\omega_c$, and the Fourier transform of this is a sharp set of lines

$$g(E)\propto\int \overline{\Psi(\boldsymbol{r}_0, t)}\,\mathrm{e}^{\mathrm{i}Et/\hbar}\,\mathrm{d}t\propto\sum_\nu \mathrm{e}^{\mathrm{i}\nu E/\hbar\omega_c}\propto\sum_n \delta(E-n\hbar\omega_c).$$

Apart from the fractional correction γ, which our slapdash treatment has managed to lose, the result is the desired quantization of levels $\hbar\omega_c$ apart.

When the argument is applied to a breakdown network such as fig. 6, the form of $\overline{\Psi(\boldsymbol{r}_0, t)}$ becomes more complex, though it may still be expected to approximate to a sequence of pulses. But now the pulses do not arrive in a simple regular sequence, but at such times as are permitted by any of the closed paths through the network, and with such amplitudes as are determined by the probabilities of reflection and transmission at each junction. The density of states therefore contains periodicities characteristic of every possible closed orbit, and it is not surprising that it takes a complicated form. Indeed, the matter is a good deal more complicated than might be guessed, since the movement of levels both up and down, like those due to $\mathscr{T}$ and $\mathscr{H}$ in the hexagonal network, and at different speeds, means that although the Fourier amplitudes in $g(E)$ may be fairly

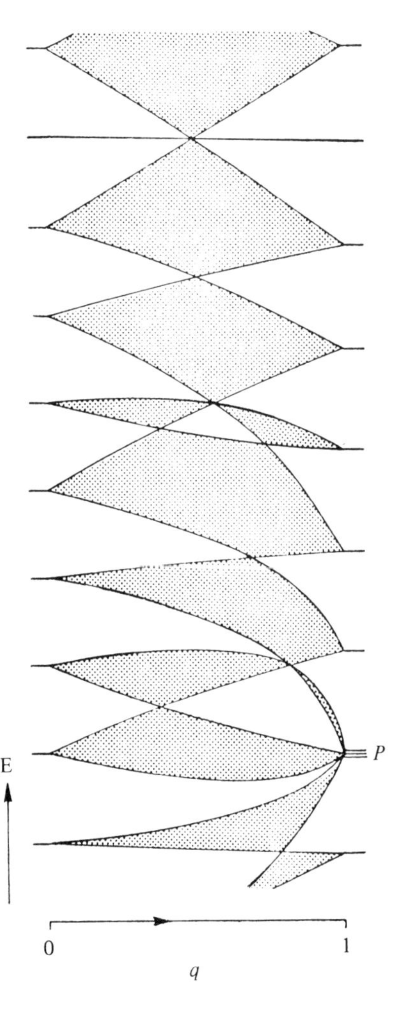

Fig. 8. Theoretical energy level structure for the network of fig. 6 (Pippard[3]). On the right, where Bragg reflection, as measured by q, is strong, the hexagonal orbit gives rise to evenly spaced sharp levels, with a second set of more widely spaced levels, having twice the degeneracy, due to the triangular orbits $\mathscr{T}$; at P levels due to $\mathscr{H}$ and $\mathscr{T}$ are shown in coincidence. On the left, when the lattice potential has been reduced to so small a value that magnetic breakdown is complete, the levels are also sharp and are characteristic of free electrons. In between there is partial reflection and partial transmission at each junction of the network, and the degenerate levels are spread into bands by the coupling of the orbits.

well defined, their relative phases develop in different ways with B, so that the resultant itself changes form with B. Such phenomena as the deHaas–van Alphen effect or the magneto-resistance oscillations are more strongly conditioned by the Fourier amplitudes than by the phases, and the extraordinary (indeed discontinuous) evolution of $g(E)$ as a function of B has few, if any, experimental consequences. Nevertheless it is worth noting, merely as a fact of nature, what variety is generated by the coupling of orbits into an extended network. As illustration of $g(E)$ in its simplest form, i.e. with B given a special value for the sake of the computation, fig. 8 shows how the discrete level structure due to $\mathscr{H}$ and $\mathscr{T}$, on the right, spreads into bands as the lattice potential is reduced and magnetic breakdown sets in; ultimately the bands collapse again to discrete free-electron levels on the left, when the lattice potential is completely removed. Falicov and Stachowiak have verified by direct computation, in a rather simpler case, that the Fourier transform of the energy level diagram conforms to the predictions of their theory.

One last point – it was assumed that the electrons continued unperturbed in their orbits for ever, but in reality they suffer scattering and other mishaps. As a result, even in the simplest case of free electrons, the succession of pulses is gradually disrupted until only noise is left in $\overline{\Psi(\boldsymbol{r}_0, t)}$. The Fourier transform of this pattern cannot give rise to sharp lines in $g(E)$ but must spread the individual levels into more or less diffuse blurs, according as there are only a few or many recognizable pulses in the sequence. Scattering results in the broadening of energy levels, and the next chapter takes up this important theme, which involves a central problem of introducing randomness and dissipation into quantum mechanics.

16 Dissipation, level broadening and radiation

After this excursion into the field of non-linear vibrators, we now return to the harmonic vibrator and take up a point which had begun to reveal **25** itself at the end of chapter 13, where we found that under the influence of a uniform, but arbitrarily time-varying, force the vibrator never forgets its initial state. If it started in the ground state, for ever afterwards its response can be described by the movement of a compact distribution in the σ-representation of equivalent classical vibrators; only the centroid of the distribution responds to the applied force. The harmonic vibrator is thus extraordinarily resistant to randomization. To be sure, if the force is not uniform, but depends on the displacement of the oscillating particle, the result just summarized is no longer true. Nevertheless, in the most important application, where an oscillator of atomic dimensions is influenced by electromagnetic vibrations, the force due to the electric field is as nearly uniform as makes no difference, since the wavelength of electromagnetic waves at a typical atomic resonant frequency is a thousand times the size of an atom. The disturbing aspect of the resistance of a vibrator to randomization is that in all theories of black-body radiation, before and after Planck, it is assumed that material oscillators and electromagnetic vibrations in a cavity will eventually share the chaotic state that allows statistical mechanics to be applied. We shall find, on looking into the matter from a consistently quantum-mechanical point of view, that the problem is not as serious as this too-classical discussion has suggested; at least it is no more serious than other fundamental questions in statistical mechanics that most physicists are content should be analysed, and if possible resolved, by mathematicians. We shall have to consider a material oscillator and a cavity vibration as a pair of coupled harmonic oscillators, whose behaviour is best described in terms of normal modes. It is the normal modes and not the individual oscillators that are quantized independently, and the recognition of this fact not only resolves the immediate difficulty but leads us on naturally to consider the quantum mechanics of energy exchange between a material oscillator and the cavity vibrations; in other words, the atomic processes involved in electromagnetic radiation and absorption. In the present chapter we shall only make a preliminary attack on the problem, mainly by the use of classical arguments where these are reasonably safe. In later chapters it will recur in various forms and the argument will be progressively refined.

Coupled harmonic vibrators

For the present purpose no deep understanding of the nature and patterns of electromagnetic vibrations in cavities is needed. It is enough to recognize that a large cavity whose walls are perfectly conducting can be stimulated, by an oscillating dipole within the cavity for instance, to resonate at a very large number of different frequencies. Each resonant mode exhibits its characteristic pattern of antinodes and nodal surfaces, and for each mode the strength of the electric field $\mathscr{E}$ at some specified point can serve to define the field strength everywhere; thus one parameter, analogous to the displacement x of a harmonic oscillator, is sufficient for the whole, while the magnetic field $\mathscr{B}$, similarly specified by a single parameter, plays a role analogous to $\dot{x}$. The energy in the cavity at any instant, being the space-integral of the energy density, $\frac{1}{2}\varepsilon_0\mathscr{E}^2+\frac{1}{2}\mathscr{B}^2/\mu_0$, can be written in a form that exactly parallels the energy, $\frac{1}{2}\mu x^2+\frac{1}{2}m\dot{x}^2$, of a material oscillator. The Schrödinger equation for the cavity mode then takes exactly the same form as (13.1), and can be cast in the reduced form (13.2) by assigning a suitable definition to ξ. Just as $\frac{1}{2}\xi^2$ is the potential energy and $\frac{1}{2}\dot{\xi}^2$, i.e., $\frac{1}{2}(\mathrm{d}\xi/\mathrm{d}\tau)^2$, the kinetic energy of the material oscillator, so $\frac{1}{2}\xi^2$ is the total cavity energy due to the electric field and $\frac{1}{2}\dot{\xi}^2$ that due to the magnetic field. In free oscillation the energy is interchanged between these two forms twice per cycle, in the same way as the energy of a harmonic vibrator is alternately potential and kinetic.

An oscillating dipole placed at some point inside the cavity, where its moment $\boldsymbol{p}$ is acted upon by the electric field $\mathscr{E}$ at that point, is coupled to the cavity mode by a coupling energy $\boldsymbol{p}\cdot\mathscr{E}$; this is proportional to $\xi_1\xi_2$ if ξ_2 is a measure of the electric field in the cavity and ξ_1 of the dipole moment. Dipole and cavity mode together behave like a pair of harmonic vibrators with potential coupling, and the classical and quantal analyses can be carried out for such a coupled pair without considering the actual nature of the vibrators.

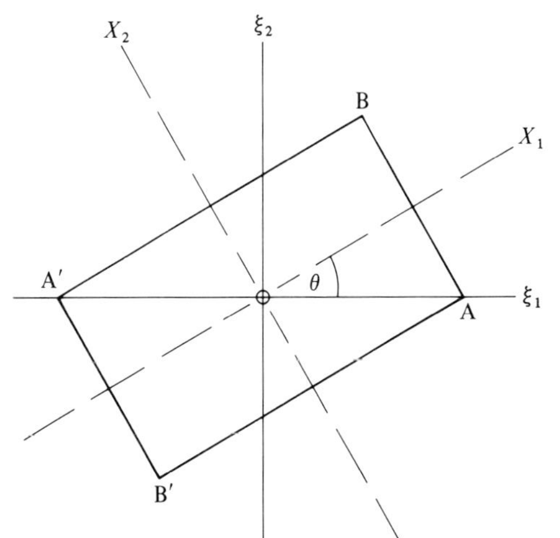

Fig. 1. Co-ordinates used to describe normal modes of two coupled harmonic vibrators (cf. fig. 12.3); ξ_1 and ξ_2 are co-ordinates of the individual vibrators, but when they are coupled the normal modes are represented by independent harmonic vibrations, at different frequencies, along X_1 and X_2.

I.366 There is no need to recapitulate the classical theory of coupled harmonic vibrators which was developed in chapter 12. The essential result is that if co-ordinates ξ_1 and ξ_2 are used, such that $\frac{1}{2}\dot{\xi}_{1,2}^2$ are the contributions to the kinetic energy of the uncoupled vibrators, then when they are coupled by a potential energy term proportional to $\xi_1\xi_2$ the normal modes are linear combinations obtained by rotating the axes in the (ξ_1, ξ_2)-plane, as in fig. 1 which differs from fig. 12.3 only in notation. Formally, the Hamiltonian for the coupled material vibrators takes the form

$$\mathcal{H}=p_1^2/2m_1+p_2^2/2m_2+\tfrac{1}{2}m_1\omega_1^2x_1^2+\tfrac{1}{2}m_2\omega_2^2x_2^2+Cx_1x_2. \tag{1}$$

If ω_1 and ω_2 are almost the same it is appropriate to define $\xi_{1,2}$ as $(m_{1,2}\bar{\omega}/\hbar)^{\frac{1}{2}}x_{1,2}$ and τ as $\bar{\omega}t$, where $\bar{\omega}=\frac{1}{2}(\omega_1+\omega_2)$. Then

$$\mathcal{H}=\tfrac{1}{2}\hbar\bar{\omega}(\dot{\xi}_1^2+\dot{\xi}_2^2+\omega_1^2\xi_1^2/\bar{\omega}^2+\omega_2^2\xi_2^2/\bar{\omega}^2+4\kappa\xi_1\xi_2/\bar{\omega}),$$

where $\kappa=C/[2\bar{\omega}(m_1m_2)^{\frac{1}{2}}]$. Writing $\xi_1=X_1\cos\theta+X_2\sin\theta$, $\xi_2=X_2\cos\theta-X_1\sin\theta$, such that

$$\tan 2\theta=2\kappa/(\omega_2-\omega_1), \tag{2}$$

we have that

$$\mathcal{H}=\tfrac{1}{2}\hbar\bar{\omega}(\dot{X}_1^2+\dot{X}_2^2+\Omega_1^2X_1^2|\bar{\omega}^2+\Omega_2^2X_2^2|\bar{\omega}^2), \tag{3}$$

the cross term having been eliminated. The normal mode frequencies, Ω_1 and Ω_2, are the solutions of the equation already derived as (12.11),

$$(\Omega_{1,2}-\bar{\omega})^2=\kappa^2+(\Delta\omega)^2, \tag{4}$$

in which $\Delta\omega=\frac{1}{2}(\omega_1-\omega_2)$. By working in normal mode co-ordinates the Schrödinger equation corresponding to (3) is made separable, and $\Psi(X_1,X_2)=\Psi_1(X_1)\Psi_2(X_2)$, Ψ_1 and Ψ_2 being solutions of harmonic oscillator equations for particles of the same mass but subject to different force constants. Alternatively (3) may be taken as the Hamiltonian of a particle moving on a plane in an anisotropic quadratic potential well. Its free motion, as a classical particle, would be a Lissajous figure traversed in the beat period, $2\pi/(\Omega_1-\Omega_2)$. As a quantized system, the component wave-functions would develop independently as already discussed, while the centroid of $\Psi\Psi^*$ would execute the classical Lissajous figure.

One can now see how randomization may enter. Let ξ_2 be the co-ordinate describing a cavity mode, and ξ_1 the dipole coupled to it. At the start of the process let us suppose both are in their ground state, so that a nearly circular Gaussian wave-packet sits at the origin in fig. 1. Excitation of the dipole by an impulsive electric field starts the wave-packet oscillating, unchanged in form, along the ξ_1-axis, say between A and A′; the subsequent Lissajous figure covers the rectangle ABA′B′, and the cavity oscillation is described by a Gaussian wave-packet whose amplitude of vibration slowly fluctuates at the beat period. So far no random element has appeared. Now let us imagine the dipole, uncoupled for the moment, to be perturbed by collisions with gas molecules; these interactions are strongly position-dependent and it is not unreasonable to suppose that the wave-function for the dipole is so disrupted that ultimately it is best described as the superposition of eigenstates whose amplitudes and phases are random. When the same process takes place while the dipole is coupled to a cavity vibration the randomization must include both. For any disturbance to the ξ_1 part of a normal-mode wave-function mixes in other normal-mode wave-functions, and therefore changes the ξ_2 part. The difference between this new treatment, which allows the transfer of randomness, and the old, which did not, lies in keeping both X_1 and X_2 as independent co-ordinates in (1) and thus in the Schrödinger equation; previously the equation for X_2 (or ξ_2) involved X_1 as a dependent variable, $X_1(t)$ defining the force that perturbed X_2. It is clear from this discussion that it is not enough to know how $\langle X_1 \rangle$ changes, or even the probability distribution for all possible changes of $\langle X_1 \rangle$; nothing of this sort can generate randomness in X_2. The mechanism of randomness is essentially non-classical, and this discussion serves to demarcate rather clearly the limit of classical reasoning about a harmonic oscillator – as soon as it interacts with other quantized systems, even another harmonic oscillator, great caution is needed if one hopes to avoid a fully quantal treatment of the coupled system. The argument applies, of course, equally well to the collision of a molecule with the oscillator; we implied above that the character of the interaction was in itself enough to produce a random outcome, but quantum mechanics makes this all the more inevitable.

Dissipation and level broadening

The foregoing analysis provides the basis for discussing the quantum mechanics of a vibrator subject to dissipation. The aim of this section is to clarify the meaning of the term *level broadening*. A classical lossy vibrator shows a Lorentzian resonance peak as its response to a force that may not be exactly tuned in frequency to the vibrator, and the quantal analogue is a slightly imprecise energy level which allows transitions to occur with emission or absorption of a quantum lying within a band of energy. In classical mechanics it is frequently permissible to introduce a loss term into the equation of motion without specifying its origin; thus a frictional force may be stated to be independent of velocity without a full treatment of the surface phenomena responsible for friction. The example we have just discussed, however, of a coupled dipole and cavity mode, illustrates how essential it is in quantum mechanics to be aware of the microscopic mechanisms at work, and to incorporate them in the analysis if necessary. There is no way of inserting rubbing friction into Schrödinger's equation, short of the prohibitively complicated expedient of writing down the many-body equation for every atom involved. We therefore base our discussion on realistic models of what may lead to dissipation in the microscopic world, and we must distinguish different types of dissipation, as well as recognize

I.136

that not all apparent level broadening is the result of dissipation. The following three examples will make the distinctions clear:

(1) *A particle in a potential well, through whose walls it may tunnel.* The solution of Schrödinger's equation in this case is virtually identical with the solution of a transmission line problem in which a resonator is defined by two reactances connected across the line, allowing weak transmission to the lines outside of the standing wave confined between them. It makes little difference, when the loss rate is slow, whether it arises from imperfect reflection of a wave incident on the barrier or by a process of absorption distributed throughout the well. The latter model crudely simulates the bound state of an unstable particle, or of one that by interaction with its surroundings may be scattered into a state irrelevant to the problem. The details of such processes should strictly be included in the treatment, but it is sometimes adequate to introduce particle non-conservation through the device of a complex potential. A particle constrained to move on a section of a line, of length a, on which the potential is $-\mathrm{i}V''$, is formally governed by Schrödinger's equation:

$$(\hbar^2/2m)\,\mathrm{d}^2\psi|\mathrm{d}x^2+(E+\mathrm{i}V'')\psi=0$$

and if $\psi=0$ at the two ends, $x=0$ and a, the nth eigenstate, $\psi_n=\sin(n\pi x/a)$, demands that $E_n=n^2\pi^2\hbar^2/(2ma^2)-\mathrm{i}V''$. The full wave-function, $\Psi_n=\psi_n\,\mathrm{e}^{-\mathrm{i}E_nt/\hbar}$, therefore has a decrement $\mathrm{e}^{-V''t/\hbar}$. The impulse-response function of this system, corresponding to the admixture after the impulse of other eigenstates, typically (from such products as $\Psi_m^*\Psi_n$) contains terms varying as $\mathrm{e}^{-\mathrm{i}(E_n-E_m^*)t/\hbar}$, an oscillation at the beat frequency but decaying as $\mathrm{e}^{-2V''t/\hbar}$. Correspondingly the frequency-response shows a Lorentzian resonance of half-width $2V''/\hbar$.

(2) *Doppler broadening of an otherwise sharp spectral line.* Here we have an ensemble of atoms moving at different speeds but all radiating identically from the point of view of an observer moving with the atom. The broadened line observed in a spectrometer is really composed of a multitude of sharp lines, unequally shifted by the relative motion, and does not constitute an example of line-broadening in the sense we are concerned with.

(3) *Radiative broadening.* This is the case to concentrate on, representing as it does a vibrating particle which, in contrast to the first example, is itself conserved but which is losing energy by radiation. We shall start with a strictly classical analysis, by a process which allows plausible extension to quantum mechanics.

The standard treatment of a Hertzian dipole[1] allows it to radiate into free space, but there are certain difficulties about incorporating unlimited free space into Schrödinger's equation, and we shall therefore return to the system already discussed, where a dipole is placed inside a large cavity, whose size may ultimately be increased without limit. There is now, to be sure, no mechanism of energy loss since the decay of the dipole oscillation accompanies the transfer of its energy to a huge number of loss-free normal modes of electromagnetic vibration in the cavity. It is possible that from time to time some of this energy will be returned to the dipole, but normally this will not be a perceptible effect, unless one makes the error of siting the dipole at the centre of a spherical cavity, when the radiated wave will be reflected and refocussed on the dipole. Even this is of small importance if the cavity is so large that the reflected wave only returns long after the dipole has radiated the greater part of its energy. And for this reason we need not worry about the inevitability of such a process taking place in the simplified model we shall treat first, in which the oscillating dipole is represented by a resonant circuit and the cavity is replaced by a one-dimensional analogue, a long transmission line, as in fig. 2.

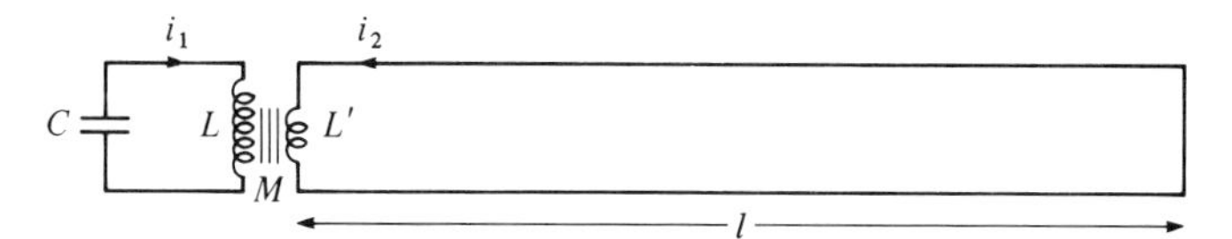

Fig. 2. Model of a radiating harmonic vibrator in the form of an *LC* circuit inductively coupled to a very long, lossless transmission line.

The line alone, uncoupled to the resonant circuit, behaves like one that is short-circuited at both ends, if L' is negligibly small. It has a closely spaced series of normal modes in which there is an integral number of half-wavelength standing wave loops; the nth mode has a frequency $n\pi c/l$, and the spacing is constant, $\pi c/l$. Hence the density of states $g(\omega)$, the number of modes in unit range of ω, is also constant with a value $l/\pi c$. When the circuit is coupled to the line the normal modes combine excitation of both line and circuit, and the coupling is especially evident for those modes whose frequency lies near the resonant frequency, ω_0, of the uncoupled circuit. As a result, the normal mode frequencies are displaced from their evenly spaced sequence, and the density of states is distorted. The first stage of the calculation is to show that a Lorentzian peak, centred on ω_0, is superimposed on the originally flat curve for $g(\omega)$.

Seen from the line, the resonant circuit presents an impedance Z_r, where

$$Z_r = \omega^2 M^2/(\mathrm{j}\omega L - \mathrm{j}/\omega C) \sim \omega_0^2 M^2/2\mathrm{j}(\omega - \omega_0)L$$

when ω is very near ω_0.† The same Z_r would alternatively be produced by extending the length of the line by Δl, if $\mathrm{j}Z_0 \tan(\omega \Delta l/c)$ is made equal to Z_r. The circuit therefore changes the frequency of any given mode by $\Delta\omega$, where $\Delta\omega/\omega = -\Delta l/l$; **I.165**

i.e.
$$\Delta\omega = (c/l)\tan^{-1}(\delta_0/\delta), \tag{5}$$

in which δ is written for $\omega - \omega_0$ and δ_0 for $\omega_0^2 M^2/2Z_0 L$. Modes well above and well below ω_0, for which $|\delta_0/\delta| \ll 1$, seem to be hardly affected, but this is a misleading interpretation. As δ goes from large positive to large negative, $\tan^{-1}(\delta_0/\delta)$ changes by π, so that on one side or other of the resonance the mode pattern slips by the mode spacing, to allow one extra

† For this circuit analysis, the electrical convention is adopted, as explained in chapter 3; j may be read as −i, if desired. **I.52**

mode to appear, being that contributed by the resonant circuit. The rearrangement principally takes place in a frequency range of the order of δ_0, and if the line is so long that $c/l \ll \delta_0$ any shift is much smaller than the frequency range of interest. It is then permissible to treat the modes as forming virtually a continuum with a density $g(\omega)$ varying with δ; instead of the original value $l/\pi c$, the modified density of states has the form

$$g(\omega) = (l/\pi c)\left(1 - \frac{\mathrm{d}}{\mathrm{d}\delta}(\Delta\omega)\right) = l/\pi c + (\delta_0/\pi)/(\delta_0^2 + \delta^2), \qquad (6)$$

from (5). The extra density of states appears as the second term at the right of (6), a Lorentzian peak of unit area, and width $2\delta_0$ at half-power, with a Q-value of $\omega_0/2\delta_0$.

The width of this peak is governed by the rate of radiation of energy from the resonant circuit into the transmission line. There is no doubt about what this rate is, since circuit theory allows the infinite line to be replaced by a resistance Z_0, appearing as ω^2M^2/Z_0 in the resonant circuit. Now the oscillation in an LCR circuit decays as $\mathrm{e}^{-\omega'' t}$, with ω'' equal to $R/2L$; hence $\omega'' = \delta_0$ and Q for the circuit is $\omega_0/2\delta_0$, the same as characterizes the Lorentzian peak in the density of states. This argument, however, employs an infinite line and is not conveniently translated into quantal terms. Let us therefore derive the result by an alternative, if more roundabout, approach involving the transfer of energy to individual modes of the finite lossless line. The central idea to be brought out, and made quantitative, is that a state of affairs in which the circuit is excited into oscillation, while the transmission line is unexcited, is not a normal mode of the coupled system. Indeed its representation in terms of normal modes involves all of the latter, and it is the beating of these against each other that leads to the exponential decay of the circuit oscillation.

I.32

A synthetic approach to this question is the simplest; we write down a suitable combination of normal modes and then demonstrate that it shows the required properties. In this case it describes a sharp pulse incident on the unexcited resonant circuit and reflected as a pulse, but trailing behind it an exponentially damped wave that is immediately attributable to radiation from the circuit, following excitation by the pulse. A typical normal mode of amplitude A_n and phase ϕ_n has a standing wave pattern for $\mathscr{E}$, the field strength on the line:

$$\mathscr{E}(x, t) = A_n\, \mathrm{e}^{\mathrm{j}(\omega_n t + \phi_n)} \sin(\omega_n x/c - \theta_n), \qquad (7)$$

in which ω_n must take the value $(n\pi + \theta_n)c/l$ in order that $\mathscr{E}$ shall vanish at the far end. At the near end $\mathscr{E}$ does not vanish at $x = 0$ but at $x = -\Delta l$ on account of the impedance presented by the circuit. Hence, from (5), $\tan\theta_n = \delta_0/\delta_n$.

The particular synthesis needed for the present purpose is that in which A_n is independent of n, and $\phi_n = \theta_n$, so that apart from a constant multiplier, which may be ignored,

$$\mathscr{E}(x, t) = \sum_n \{e^{j\omega_n(t+x/c)} - e^{j\omega_n(t-x/c)+2j\theta_n}\}.$$

Since the modes are almost equally spaced in frequency, summation may be replaced by unweighted integration; and the resultant variations of $\mathscr{E}(x, t)$ are expressed in the form

$$\mathscr{E}(x, t) = \int_0^\infty [e^{j\omega(t+x/c)} - e^{j\omega(t-x/c)+2j\theta(\omega)}]\, d\omega. \tag{8}$$

The first term describes a sharp pulse located at $x = -ct$ travelling towards the left-hand end and reaching it at time $t = 0$; thenceforth it lies to the left and has vanished from the picture. The presence of $\theta(\omega)$ in the second term, which describes the subsequent behaviour at positive x, shows that the reflected pulse is distorted. Let us evaluate this second term at the point $x = 0$, writing it in the form:

$$\mathscr{E}(0, t) = -e^{j\omega_0 t} \int_{-\infty}^{\infty} e^{j[\delta t + 2\tan^{-1}(\delta_0/\delta)]}\, d\delta, \tag{9}$$

in which $\delta = \omega - \omega_0$, as before. The lower bound has been extended from $-\omega_0$ to $-\infty$, a change which facilitates evaluation of (9) by contour integration without damaging the interesting part of the result which is the after-effect that follows the pulse. The integrand in (9) has one simple pole at $\delta = j\delta_0$. When $t < 0$ the term $e^{j\delta t}$ decreases exponentially as one proceeds in the negative imaginary direction of δ, and the negative semicircle at infinity makes no contribution to the contour integral which also includes the real axis. Since no poles are enclosed (9) vanishes at all negative t. When t is positive, however, the positive semicircle must be used to complete the contour, which now includes the pole. Thus if we disregard a constant factor in the residue, the integral is governed by the value of $e^{j\delta t}$ at the pole, i.e., $e^{-\delta_0 t}$, and $\mathscr{E}(0, t) \propto e^{(j\omega_0 - \delta_0)t}$ when $t > 0$. This decaying oscillation of $\mathscr{E}$ is of course induced by the circuit oscillation, and the analysis shows that it decays as $e^{-\delta_0 t}$, the result already known from circuit theory.

To conclude this section, it is worth stressing that the Lorentzian peak of width $2\delta_0$ is a peak in the density of very closely spaced sharp lines, each a loss-free normal mode of the complete system. If one likes to think of it as an extra, broadened, line characteristic of the radiation-damped oscillator alone, superimposed on the regular continuum of transmission line modes, no harm is done and it is probably a useful pictorial simplification. But fundamentally a broadened energy level has no meaning in quantum mechanics, and it is wise to bear in mind the interpretation in terms of a broad peak in the density of states.

Electromagnetic radiation into free space

It is easy to extend the argument to the radiation of energy by an oscillating dipole into the modes of a large cavity.† Care is needed, however, in specifying the coupling of the dipole to the cavity modes. It must be remembered that of the many modes lying within any narrow frequency range some will have nodes and some antinodes of the standing waves of electric field $\mathscr{E}$ at the location of the dipole. The former will be uncoupled and the latter most strongly coupled, and all the intermediate cases will occur. Moreover all polarizations of $\mathscr{E}$ will be represented but only the component parallel to the dipole will be coupled to it. These effects are included in the argument by assigning the same, i.e. the average, coupling to each mode, simply by assuming the mean square value of the field responsible for coupling to be $\frac{1}{3}\bar{\mathscr{E}}^2$, where $\bar{\mathscr{E}}^2$ is the true mean square field in the cavity.

We now express δ_0 in the transmission line example in a form that allows its analogue in the cavity to be written down immediately:

$E_1 \equiv$ energy in oscillating circuit $= \frac{1}{2}Li_1^2$

$E_2 \equiv$ energy in a transmission line mode $= \frac{1}{4}Z_0 l i_2^2/c$

$E_{12} \equiv$ coupling energy $= Mi_1 i_2$

$g(\omega) \equiv$ density of modes at circuit frequency $= l/\pi c$

The currents i_1 and i_2 are peak values in the circuit and on the line. In terms of these variables,

$$\delta_0 = \omega_0^2 M^2/2Z_0 L = \pi\omega_0^2 g(\omega_0)E_{12}^2/16E_1E_2. \tag{10}$$

Now for a dipole consisting of a mass m, carrying charge e and oscillating with amplitude a, the corresponding quantities for a given cavity mode are as follows:

$$E_1 = \tfrac{1}{2}m\omega_0^2 a^2, \qquad E_2 = \tfrac{1}{2}\varepsilon_0\bar{\mathscr{E}}^2 V = \tfrac{1}{2}\bar{\mathscr{E}}^2 V/\mu_0 c^2 \quad \text{and} \quad E_{12} = ea\mathscr{E}_\parallel,$$

where V is the volume of the cavity and $\mathscr{E}_\parallel$ is the peak electric field parallel to the dipole (we have not yet taken the average). Furthermore[2] $g(\omega_0) = \omega_0^2 V/\pi^2 c^3$. Substituting in (10), and writing $e^2a^2\bar{\mathscr{E}}_\parallel^2 = \frac{1}{3}e^2a^2\bar{\mathscr{E}}^2$ for the average value of E_{12}^2, we have

$$\delta_0 = \mu_0\omega_0^2 e^2/12\pi mc.$$

The time-constant for energy decay, τ_e, is $1/2\delta_0$;

i.e.
$$\tau_e = 6\pi mc/\mu_0\omega_0^2 e^2. \tag{11}$$

Alternatively expressed, the rate of energy loss by the dipole,

$$-\dot{W} = E_1/\tau_e = \mu_0\omega_0^4 e^2 a^2/12\pi c, \tag{12}$$

† Easy, but not obviously justifiable. A very serious problem of convergence arises here which is not present in the previous example. This will be discussed at the end of the chapter; it is a pervasive difficulty of electromagnetism but here, as in many other cases, it can apparently be ignored without invalidating the theory.

81

which is just Hertz's[(1)] expression for the rate of radiation by a dipole into free space. We have therefore satisfied ourselves that the dissipative radiative process can be represented by a multitude of non-dissipative energy exchanges with the normal modes of a cavity, provided the cavity is taken to be large enough for the spectrum of modes to be treated as a continuum. This is the same criterion as ensures that no reflected wave returns to the dipole until long after the decay is complete. Then at any instant the reaction of the cavity on the dipole is the same as if the cavity were completely unexcited until that very moment; since the system is linear and the rate of decay is always governed only by the instantaneous condition, it cannot be other than exponential. This argument carries over into the quantal treatment, and we shall take it for granted from now on. Thus, if we find the initial rate of energy loss immediately after coupling to the cavity to be αE_1, we shall infer a time-constant $1/\alpha$ for energy dissipation.

To illustrate this point, let us return to the transmission line model of fig. 2 and consider how the energy in the nth mode varies with time. At a frequency $\omega_n + \Delta$ the line presents at the input end an inductance $Z_0 \tan[(\omega_n + \Delta)l/c]$, i.e. $\Delta Z_0 l/c$ for small Δ. The same inductance would be presented by an LC circuit, resonant at ω_n, if $L = \frac{1}{2}Z_0 l/c$; it is convenient to replace the transmission line mode by this lumped circuit, so that the problem of energy transfer to the mode is reduced to the standard problem of energy transfer between coupled circuits. For the purpose of determining the initial decay of E_1 we assume the current i_1 to oscillate with constant amplitude and to excite the coupled LC circuit with an emf $\omega_0 M i_1 \sin \omega_0 t$. The response after switching on the excitation at $t = 0$ takes the form (see 6.31): **24**

$$i_2(t) = -i_2 \sin \bar{\omega} t, \tag{13}$$

in which

$$\bar{\omega} = \tfrac{1}{2}(\omega_0 + \omega_n),$$

$$i_2 = (2\omega_0 M c i_1 / Z_0 l) \sin(\tfrac{1}{2}\delta_n t)/\delta_n,$$

and

$$\delta_n = \omega_n - \omega_0.$$

The energy E_2 in the mode is $\frac{1}{4}Z_0 l i_2^2/c$;

i.e.

$$E_2 = (\omega_0^2 M^2 c i_1^2 / Z_0 l) \sin^2(\tfrac{1}{2}\delta_n t)/\delta_n^2$$

$$= (4\delta_0 E_1 / g\pi) \sin^2(\tfrac{1}{2}\delta_n t)/\delta_n^2. \tag{14}$$

By writing the result in the form (14) it has been rendered independent of details of the particular model, and can be applied to other systems. So long as $\delta_n t \ll 1$, E_2 rises quadratically with time, but sooner or later the energy is restored to the first circuit, and from then on oscillates between the two with zero mean rate of increase. When all the other modes are considered to be similarly engaged, one sees that the number still involved in the quadratic increase are those in a frequency range $\pm\delta$ that decreases steadily as $1/t$, and the product of these two time variations causes the total energy in the modes to increase as t. From this the initial rate of decay of E_1 follows, and hence the time constant for the exponential radiation damping. To derive the result in detail, it is only necessary to integrate (14) over all modes:

$$E_1(0) - E_1(t) = \int_{-\infty}^{\infty} \dot{E}_2 g \, \mathrm{d}\delta_n = (2\delta_0 E_1 t/\pi) \int_{-\infty}^{\infty} \sin^2 x \, \mathrm{d}x/x^2 = 2\delta_0 E_1 t, \tag{15}$$

equivalent to a time constant $\tau_e = 1/2\delta_0$. This rate can be obtained by neglecting all the modes except those in a frequency range $2\pi/t$, for each of which perfect tuning is assumed, with $\delta_n = 0$. Thus each such mode makes a contribution to the initial decay of E_1 in the form (14) with $\delta_n = 0$:

$$E_1(0) - E_1(t) = \delta_0 E_1(0) t^2/g\pi = E_1(0) t^2/2\pi g \tau_e, \tag{16}$$

and the effective number of modes is $2\pi g/t$; the product yields (15).

Spontaneous radiation in quantum mechanics

We must now show how the essential features of these arguments can be carried over into quantum mechanics, and for this purpose we return to the basic model of an oscillating dipole interacting with a single cavity mode, the whole being represented by two coupled loss-free vibrators. As already noted, as Ψ evolves on the plane of fig. 1 the centroid $\langle \boldsymbol{r} \rangle$, where $\boldsymbol{r}$ represents the co-ordinates (ξ_1, ξ_2), executes a Lissajous figure of classical form. One cannot infer from this, however, that the energy exchange between the vibrators also behaves classically. The response of $\langle r^2 \rangle$ to an applied force may follow classical rules, but we have already seen that adequate discussion of the energy transfer between quantized systems demands that the complete system be treated as a single quantized unit.

67

To appreciate what rules govern the exchange of energy let us first note some general points. If at any instant the wave-function for the two coupled vibrators, expressed in terms of normal mode co-ordinates, is $\Psi(X_1, X_2, t)$, this can be Fourier-analysed in terms of eigenfunctions which are themselves products of normal mode oscillator functions:

$$\Psi(X_1, X_2, t) = \sum_n \sum_m a_{nm} \psi_n(X_1) \psi_m(X_2) \, \mathrm{e}^{-\mathrm{i}[(n+\frac{1}{2})\Omega_1 + (m+\frac{1}{2})\Omega_2]t}, \tag{17}$$

in which Ω_1 and Ω_2 are the normal mode frequencies, assumed to be fairly close together. If many terms are present in (17) the different frequencies are responsible for a complicated evolution of Ψ, but however complicated it may be the amplitude of Ψ must be strictly periodic at the beat frequency $\chi = |\Omega_1 - \Omega_2|$. Different physical quantities would show on evaluation different periodic patterns of behaviour, according to the harmonics of the beat frequency that appeared. The energy in either of the constituent vibrators, however, behaves in the simplest way possible, with only the fundamental playing a part. To evaluate E_1, for example, we must know

$\langle\xi_1^2\rangle$ and $\langle\dot{\xi}_1^2\rangle$; the point is sufficiently illustrated by considering $\langle\xi_1^2\rangle$, which in terms of normal modes is $\langle X_1^2 \cos^2\theta - 2X_1X_2\cos\theta\sin\theta + X_2^2\sin^2\theta\rangle$. Only the cross term in X_1X_2 generates the beat frequency χ or its harmonics when this is evaluated by use of (17):

$$\langle X_1X_2\rangle = \sum_n\sum_m\sum_{n'}\sum_{m'}\iint a_{nm}a^*_{n'm'}\psi_n(X_1)\psi_m(X_2)X_1X_2\psi^*_{n'}(X_1)\psi^*_{m'}(X_2)$$
$$\times\, e^{i[(n-n')\Omega_1+(m-m')\Omega_2]t}\,dX_1\,dX_2. \tag{18}$$

Now $\int\psi_nX_1\psi'_n\,dX_1$ vanishes unless $n-n'=\pm 1$, and similarly for X_2, so that the only frequencies to emerge from (18) are $\pm\Omega_1\pm\Omega_2$, and the only low frequency is χ. The energy exchange between the two vibrators is therefore purely sinusoidal.

Now in general if the two vibrators are set up in arbitrary non-stationary states and then coupled, this sinusoidal exchange of energy will not be caught at an extremal point, i.e. $\dot{E}_1 = -\dot{E}_2 \neq 0$ initially. There are, however, certain situations in which initially $\dot{E}_1 = 0$, as for example when the vibrators are both in stationary states before coupling. More significant for the present discussion is for the cavity mode to be in its ground state while the dipole is arbitrarily excited. If, after coupling, E_2 is to oscillate at the beat frequency it must start at the minimum for the cycle, since there is no possibility that sampling at any later stage will reveal it to have an energy content less than its zero-point energy. To determine the range of oscillation of E_1 and E_2 we shall, for the sake of future reference, present the argument in a more general form than is needed for the immediate purpose.

At the moment of coupling, let the two vibrators have energies E_{10} and E_{20}:

$$E_{10} = \langle\xi_1^2 + \dot{\xi}_1^2\rangle_0, \qquad E_{20} = \langle\xi_2^2 + \dot{\xi}_2^2\rangle_0, \tag{19}$$

as follows from (13.11), if $\frac{1}{2}\hbar\omega_0$ is taken as the unit energy and all frequencies are close enough together for distinctions to be ignored, except when the beating effect is considered. Then the energies in the normal modes follow from rotating the co-ordinate axes, as in fig. 1:

$$\left.\begin{aligned} E_{n1} &= \langle X_1^2 + \dot{X}_1^2\rangle = E_{10}\cos^2\theta + E_{20}\sin^2\theta + \langle\xi_{10}\xi_{20} + \dot{\xi}_{10}\dot{\xi}_{20}\rangle\sin 2\theta \\ E_{n2} &= \langle X_2^2 + \dot{X}_2^2\rangle = E_{10}\sin^2\theta + E_{20}\cos^2\theta - \langle\xi_{10}\xi_{20} + \dot{\xi}_{10}\dot{\xi}_{20}\rangle\sin 2\theta. \end{aligned}\right\} \tag{20}$$

Now let the cavity mode be initially in its ground state, so that E_{10} and E_{20} are extrema of E_1 and E_2. Then $\langle\xi_{20}\rangle$ and $\langle\dot{\xi}_{20}\rangle$ both vanish, and the third terms in (20) are absent. After coupling, E_{n1} and E_{n2} stay constant but the two normal modes have different frequencies, and consequently E_1 and E_2 oscillate about an easily determined mean. For on applying the reverse transformation we have that

$$\left.\begin{aligned} E_1 &= E_{n1}\cos^2\theta + E_{n2}\sin^2\theta - \langle X_1X_2 + \dot{X}_1\dot{X}_2\rangle\sin 2\theta, \\ E_2 &= E_{n1}\sin^2\theta + E_{n2}\cos^2\theta + \langle X_1X_2 + \dot{X}_1\dot{X}_2\rangle\sin 2\theta, \end{aligned}\right\} \tag{21}$$

and the third terms oscillate with the beat frequency about a mean value of zero. Hence E_1 oscillates about a mean value:

$$(E_1)_{Av} = E_{n1}\cos^2\theta + E_{n2}\sin^2\theta = E_{10} + \tfrac{1}{2}(E_{20} - E_{10})\sin^2 2\theta \qquad (22)$$

from (20). Similarly

$$(E_2)_{Av} = E_{n1}\sin^2\theta + E_{n2}\cos^2\theta = E_{20} + \tfrac{1}{2}(E_{10} - E_{20})\sin^2 2\theta. \qquad (23)$$

Since E_{10} and E_{20} are extremal values for the oscillations of E_1 and E_2, it follows that E_1 oscillates between E_{10} and $E_{10}\cos^2 2\theta + E_{20}\sin^2 2\theta$, and E_2 between E_{20} and $E_{20}\cos^2 2\theta + E_{10}\sin^2 2\theta$.

This result may be compared with the beating behaviour of two classical vibrators. If at the moment of coupling, one is at rest with $E_{20} = 0$, the motion is initially represented on fig. 1 by a vibration between A and A′. After half a beat period the oscillation lies along BB′; E_2 has reached its maximum of $E_{10}\sin^2 2\theta$ and E_1 its minimum of $E_{10}\cos^2 2\theta$. The quantal behaviour is almost the same, except for the zero-point energy E_{20}; in fact the oscillations of energy are just as if the two initial values E_{10} and E_{20} indulged in independent beating in the classical manner, so that as a fraction $E_{10}\sin^2 2\theta$ was transferred from the first to the second vibrator, simultaneously $E_{20}\sin^2 2\theta$ was transferred in the reverse direction. Alternatively we may imagine only the excitation energy, $E_{10} - E_{20}$, as involved in an otherwise classical process. From this it is an obvious step to take over the whole classical description of the decay of a dipole oscillator in an empty cavity; it is not the total energy which is radiated to give exponential decay with time-constant $1/2\delta_0$, but only the excitation energy, $E - \frac{1}{2}\hbar\omega_0$, and the time-constant is the same as is given by the Hertzian theory of radiation.

Planck's radiation law; Einstein coefficients

The idea of independent energy exchange between two vibrators might be considered the most primitive form of Prévost's theory of exchanges, which is fundamental to the theory of heat transfer by radiation. The non-interaction of the traffic in both directions is readily justified in classical processes by the assumption of a random phase relationship between the bodies concerned. Thus when a vibrating dipole is placed in a cavity which is already excited, the initial power transfer results from the oscillating current of the dipole interacting with the oscillating electric field of the cavity modes; the mean value is zero if an average is taken over all possible phase relationships. It is only when each has had time to influence the vibration of the other, and introduce correlation into the phase relationship, that the mean power transfer is non-zero, and in fact is the same as if each were coupled to an unexcited vibrator. The proof of these statements is left as a simple exercise. The same is true when two quantized vibrators are coupled, and averages taken over every conceivable phase difference involved. Perhaps this is a fair representation of an assembly of dipoles placed in an excited cavity, but in fact it is not necessary to discard all phase relationships so wildly in order to achieve the effect of randomization. If the dipole and a cavity mode are established in some states of excitation, not necessarily stationary states, and are then coupled, in general there will be an immediate power transfer, with $\dot{E}_1 = -\dot{E}_2 \neq 0$ as already remarked. Let the experiment be repeated many times, however, with exactly the same initial conditions, but with varying intervals allowed to elapse before coupling, so that the relative phases of the two vibrators assume different initial values; it will be found that on the average the energy transfers in the two directions are independent.

The proof is straightforward. In each trial E_1 starts with the same value, E_{10}, but the oscillations vary in amplitude, though they are always sinusoidal with the same frequency, χ. Consequently when the average is found, it too will be sinusoidal with frequency χ. Moreover, although $\dot{E}_1$ does not initially vanish in most trials, the mean value $\dot{\bar{E}}_{10} = 0$ since positive and negative $\dot{E}_{10}$ are equally likely. Hence the oscillations of $\bar{E}_1$ start at an extremum, and are about an average value $(\bar{E}_1)_{\text{Av}}$ which is simply the mean of $(E_1)_{\text{Av}}$ taken over all the trials. Now when the vibrators are coupled after some arbitrary lapse of time there is no preferential phase correlation between the normal mode vibrations, so that the mean values of X_1X_2 and $\dot{X}_1\dot{X}_2$ are zero. Thus $(\bar{E}_1)_{\text{Av}}$ is obtained by dropping the last term in (21) and the argument proceeds exactly as if the cavity mode had been in a stationary state, with E_1 oscillating between E_{10} and $E_{10}\cos^2\theta + E_{20}\sin^2\theta$. And, as we have seen, this is the same as if the dipole was transferring its excitation energy to an unexcited cavity mode and conversely the cavity mode were transferring its excitation energy to an unexcited dipole. We may now be satisfied that when an excited dipole is placed in an excited cavity, there are enough modes of similar frequency and random phase for the mean power transfer to each to follow the same initial pattern, along the lines of (16), as if the cavity modes were all in their ground state. And similarly the average transfer from the cavity modes to the dipole starts in the same way as if the dipole were in its ground state.

It is clear from this that if the excitation energy of the dipole matches the average excitation energy for the cavity modes of the same frequency there will be no net transfer, so that thermal equilibrium between material oscillators and cavity modes is achieved when both exhibit the same dependence of mean energy on frequency. This, of course, was Planck's conclusion based on very careful, and obviously troubled, analysis of the process of interaction, and it led him to his black-body radiation law. The mean excitation energy of a quantized harmonic oscillator, and hence of a cavity mode, is $\hbar\omega/(\mathrm{e}^{\hbar\omega/k_BT}-1)$; and since the density of states per unit volume of cavity is ω^2/π^2c^3, the energy density (not counting zero-point energy) per unit range of ω takes the now familiar form:

$$u(\omega) = (\hbar\omega^3/\pi^2c^3)/(\mathrm{e}^{\hbar\omega/k_BT}-1). \tag{24}$$

Eighty years after the discovery one may have sufficient confidence in the result to point out that Planck's careful analysis was not strictly necessary.

We have seen how the choice of energy-based co-ordinates to describe coupled oscillators removes from the argument all parameters defining their magnitude or constitution. In these terms a material oscillator and a cavity mode at the same frequency become strictly equivalent, and the condition for statistical equilibrium could not be other than that their co-ordinates and hence their energy content should be the same. But to have expected Planck to see it in this light is to ignore the difficulties of a pioneer, and from the comfort of a limousine to belittle the hardships of those who drove the road through the jungle.

In 1916 Einstein published his justly famous analysis[(3)] of the black-body radiation problem in terms of spontaneous and stimulated processes. It might appear as if two rather different mechanisms were engaged, stimulated emission and absorption on the one hand, being the interaction of cavity modes with a material oscillator, and on the other hand spontaneous emission in which superficially it looks as if the cavity modes play no part. It is clear, however, from our discussion that it is exactly the same coupling between dipole and cavity modes that governs both processes, and that (at least when the dipole is a harmonic oscillator) everything may be described quite symmetrically in terms of spontaneous emission from either dipole or cavity mode into the other. Unfortunately many texts create an air of mystery about spontaneous emission by suggesting that it lies beyond the scope of Schrödinger's wave mechanics; it is even hinted sometimes that the zero-point oscillations of the cavity are needed to initiate spontaneous emission, as it were by telling the otherwise isolated dipole that there is a suitable recipient for its energy near at hand. We have seen, however, that these difficulties evaporate once it is recognized that free vibrations of the dipole in an unexcited cavity do not constitute a normal mode.

169

As for the role of zero-point energy, what has emerged is that it is irrelevant to this discussion and is better ignored throughout – only excitations above the ground state enter the description of energy transfer.†

Nevertheless there is still a reconciliation to be effected between Einstein's vision and what has been developed here, and for this we apply his argument to an assembly of similar dipoles (harmonic oscillators) coupled to black-body radiation in a cavity. In the historical context of Bohr's quantum theory it is natural to suppose that each oscillator occupies at any one moment a well-defined energy level. Of the n_l oscillators occupying the lth level, some are making spontaneous transitions to the $(l-1)$th while some are being stimulated upwards and others downwards by the radiation. The rates, $-\dot{n}_l$, for the three processes are to be written, according to Einstein, as $A_l n_l$, $B_{l+1} n_l u(\omega_0)$ and $B_l n_l u(\omega_0)$. Only neighbouring levels are involved, and A_l and B_l are the coefficients for spontaneous and stimulated processes between l and $l-1$, while B_{l+1} is the coefficient for stimulated

† This should not be taken to imply, however, that zero-point effects are in principle unobservable; they have a deleterious influence on X-ray and neutron diffraction intensities, and have been invoked to explain the fluidity of helium down to 0 K.

processes between l and $l+1$. The mean rate of energy gain by each oscillator that happens to be excited in the lth level may therefore be written:

$$\dot{W}_l = (-A_l - B_l u + B_{l+1} u)\hbar\omega_0. \tag{25}$$

Now we have already seen that the oscillator strength is proportional to the quantum number of the upper of the two levels involved, so that it is appropriate to write B_l as βl, β taking the same value for all levels. Hence

18

$$\dot{W}_l = (-A_l + \beta u)\hbar\omega_0, \tag{26}$$

which is in exactly the form required to reconcile the different viewpoints. Each oscillator in the lth level is emitting energy at a rate $A_l\hbar\omega_0$, independent of u, while the cavity radiation is feeding it at a rate $\beta u\hbar\omega_0$, independent of l. It only remains to show that the calculated expressions for A_l and β make sense. For $A_l\hbar\omega_0$ we go back to (12), taking a to be the amplitude of a classical vibrator with energy $l\hbar\omega_0$, so that

$$A_l = \mu_0\omega_0^2 e^2 l/6\pi mc. \tag{27}$$

For β we note that $\beta u\hbar\omega_0$ is the rate of absorption of energy from the radiation field by an oscillator in its ground state, $l=0$, for which $\beta_{l+1}=\beta$, and that this is the same rate as for the equivalent classical oscillator, starting from rest. Randomness of phase causes the increase in energy due to each mode to be additive, and the same argument as led to (16) applies here, that after time t the overall effect is as if only modes in the frequency range $2\pi/t$ around ω_0 had contributed. The energy density in these modes is $2\pi u/t$, so that the mean square of the component of electric field that may be considered to act on the dipole is $\frac{4}{3}\pi u/\varepsilon_0 t$. Since a field $\mathscr{E}$ oscillating in resonance imparts energy $\frac{1}{8}e^2\mathscr{E}^2t^2/m$ in time t, it follows that the dipole gains energy at a rate $\frac{1}{6}\pi e^2 u/\varepsilon_0 m$ or $\frac{1}{6}\pi\mu_0 e^2c^2u/m$.

Hence

$$\beta = \tfrac{1}{6}\pi\mu_0 e^2c^2/\hbar m\omega_0, \tag{28}$$

and

$$A_l/B_l = \hbar\omega_0^3/\pi^2c^3 = \hbar\omega_0 g/V. \tag{29}$$

As is well known from standard treatments of this problem, (29) is the ratio of the coefficients for spontaneous to stimulated processes that is needed for Planck's radiation law to emerge from Einstein's theory.

The equation governing overall energy exchange between the cavity radiation and an assembly of oscillators follows from (26), which may be written by use of (29) in the form:

$$\dot{W}_l = \beta\hbar\omega_0(u - l\hbar\omega_0 g/V). \tag{30}$$

Summing over all oscillators we find that their total excitation energy E obeys the equation:

$$\dot{E} = \sum_l n_l\dot{W}_l = \beta\hbar\omega_0(Nu - gE/V), \tag{31}$$

since N, the total number of oscillators, is Σn_l and $E = \Sigma n_l l\hbar\omega_0$. Hence the mean energy of an oscillator, $\bar{E}(=E/N)$ relaxes towards its equilibrium value of uV/g according to the equation:

$$\dot{\bar{E}} = (\beta\hbar\omega_0 g/V)(uB/g - \bar{E}), \tag{32}$$

and the time-constant, τ_e, is $V/\beta\hbar\omega_0 g$, independent of u. This is exactly what we have been led to expect by the quasi-classical picture of cavity modes and oscillators exchanging energy as if each alone were excited. The value of τ_e is of course the same as the classical (Hertzian) value given by (11).

In placing so much emphasis on the strict reciprocity of the exchanges between a material oscillator and a cavity mode, there is some danger of hiding an important distinction between spontaneous and stimulated processes. Consider, for instance, radiation falling on a material surface and being partially reflected and partially absorbed. If the material is cold its response to the radiation is entirely to be ascribed to stimulated processes, and the reflection and absorption coefficients are controlled by the amplitude and phase of the dipole vibrations induced by the incident field. If the material is hot its own vibrators are excited quite independently of the incident field, and with no organized phase relationship to that of the field. In classical and quantum physics these processes, when they occur in harmonic vibrators, are uncoupled. One may therefore observe the spontaneous radiation of a red-hot sheet of metal without confusing it with the reflection of a bright light shone onto it; the extra dipole vibrations caused by the latter are phase-coherent with the source. It is not so easy to appreciate this point if one thinks of the material vibrators as existing only in pure quantum states, since these have no dipole moment. However, if we picture the incident radiation as possessing a well-defined phase, we **21** know from chapter 13 that the additional motion excited in the vibrators will be essentially the same as if they were classical, and initially at rest. The energy transfer from field to vibrators is thus accompanied by coherent excitation, and conversely this reacts back on the radiation field also in a coherent fashion. These are the processes of stimulated absorption and emission. The spontaneous process is in no essential sense different from stimulated emission except that the emitting system is not phase-linked to the incident field.

We have made considerable progress without the need to set up quantum-mechanical equations of motion, but to go beyond this point demands more detailed analysis. Further discussion of radiative processes will therefore **104, 164** be postponed, to be resumed in chapters 18 and 20.

Divergences in the theory of dipole radiation; mass enhancement

We discussed the level broadening resulting from spontaneous radiation in terms of the interaction of a material oscillator with the cavity modes having very similar frequencies. We did not enquire into the effect of high-frequency modes which unhappily lead to a divergent integral. It is convenient to reformulate the problem in terms of an ensemble of vibrators, each representing a cavity mode, coupled to the material oscillator but not among themselves. The normal modes of such an arrangement can be expressed as the solutions of a comparatively simple equation.

For simplicity let us ignore variations, from one mode to another, of polarization and location of the field maxima, treating all modes as coupled to the oscillator with the mean value of the coupling constant. When the energy of the ith mode is wholly electrical and of magnitude $\frac{1}{2}\varepsilon_0\bar{\mathscr{E}}^2 V$, we equate this to the energy of the equivalent material vibrator, $\frac{1}{2}\mu_i x_i^2$, and take $(\frac{1}{3}\bar{\mathscr{E}}^2)^{\frac{1}{2}}$ as the field acting on the oscillating charge. Thus the field has strength $(\mu_i/3\varepsilon_0 V)^{\frac{1}{2}} x_i$ and exerts a force $c_i x_i$ on the oscillating charge, where $c_i = e(\mu_i/3\varepsilon_0 V)^{\frac{1}{2}}$. When all modes are taken into account, the equation of motion of the material oscillator has the form:

$$m_0\ddot{x}_0 + \mu_0 x_0 - \sum_i c_i x_i = 0. \tag{33}$$

Correspondingly the dipole moment of the oscillator modifies the equation of motion of each mode in the expected reciprocal manner, as may easily be proved:

$$m_i\ddot{x}_i + \mu_i x_i - c_i x_0 = 0. \tag{34}$$

Let a normal mode solution of the set of equations consisting of (33) and the N equations analogous to (34) have frequency Ω. When this trial solution is substituted, the condition for the $N+1$ equations to be compatible is that the determinant of the coefficients shall vanish;

i.e.
$$\begin{vmatrix} \mu_0 - m_0\Omega^2 & -c_1 & -c_2 \dots \\ -c_1 & \mu_1 - m_1\Omega^2 & 0 \dots \\ -c_2 & 0 & \mu_2 - m_2^2\Omega^2 \dots \\ \vdots & \vdots & \vdots \end{vmatrix} = 0. \tag{35}$$

On multiplying out and dividing through by $(\mu_1 - m_1\Omega^2)$ $(\mu_2 - m_2\Omega^2)\dots(\mu_N - m_N\Omega^2)$ the equation for Ω is obtained in the form:

$$\mu_0 - m_0\Omega^2 = \sum_i c_i^2/(\mu_i - m_i\Omega^2), \tag{36}$$

i.e.
$$\omega_0^2 - \Omega^2 = (e^2/3m_0\varepsilon_0 V)\sum_i \omega_i^2/(\omega_i^2 - \Omega^2). \tag{37}$$

The right-hand side of (37) oscillates with great rapidity, reaching $\pm\infty$ at every value of Ω that coincides with a cavity mode. The slope of the function is least at points lying roughly midway between successive modes, as shown in fig. 3(a), and it is clear that the solutions are bunched together most strongly where the intersections of the lines representing the left and right sides of (37) lie near the positions of minimal slope. We shall take

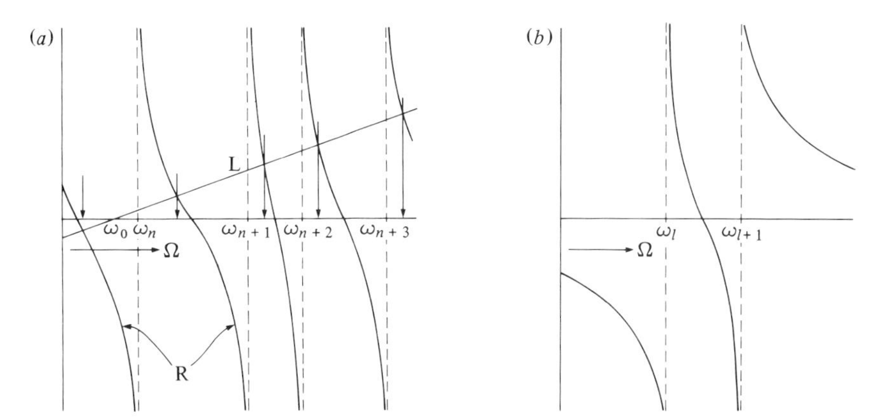

Fig. 3. (*a*) Schematic diagram of the left (L) and right (R) hand sides of (37) as functions of Ω. The normal modes are given by the intersections, as shown by the arrows, and it is the slope of L that causes the displacement of successive modes to vary progressively and leads to a slight bunching of the modes. In (*b*) is shown the contribution to R of two successive modes alone.

this as defining the centre of the broadened line. To determine its position, let us suppose it to lie between ω_l and ω_{l+1}, whose contribution to the right-hand side of (37) is shown in fig. 3(*b*). If that were all, the minimum slope would occur on the axis, and the line would have its centre at $\omega_0^2-\Omega^2=0$. But all the other modes may be expected to provide a contribution to the right-hand side of relatively low slope, but not necessarily small magnitude; the centre of the line will occur when $\omega_0^2-\Omega^2$ is equal to this contribution. We therefore evaluate (37), excluding the immediate vicinity of Ω from the sum; for large V and correspondingly dense distribution of modes, the sum may be replaced by the principal value of an integral:

$$m_0(\omega_0^2-\Omega^2)=(e^2/3\varepsilon_0)\,⨍\frac{\omega^2 g(\omega)}{(\omega^2-\Omega^2)}\,\mathrm{d}\omega, \tag{38}$$

in which $g(\omega)$ is the density of states per unit volume of cavity, i.e. ω^2/π^2c^3. Hence

$$m_0(\omega_0^2-\Omega^2)=(e^2/3\pi^2\varepsilon_0c^3)\,⨍\frac{\omega^4}{(\omega^2-\Omega^2)}\,\mathrm{d}\omega. \tag{39}$$

There is no reason to cut off the spectrum of cavity modes at any finite frequency, and the problem raised by the analysis is immediately obvious, since the integral in (39) diverges as ω^3 at high frequencies. The difficulty does not arise in the model represented in fig. 2. Analysis of inductively coupled circuits shows that the equation corresponding to (38) does not contain ω^2 in the numerator of the integral; and since for a transmission line $g(\omega)=\text{constant}$, the integral to be evaluated is $⨍_0^\infty \mathrm{d}\omega/(\omega^2-\Omega^2)$, which is zero. The earlier result is therefore confirmed, that the centre of the line is unshifted.

To return to the oscillating dipole, the integral in (39) may be conveniently rearranged

$$⨍\frac{\omega^4}{(\omega^2-\Omega^2)}\,\mathrm{d}\omega=\int(\omega^2+\Omega^2)\,\mathrm{d}\omega+\Omega^4⨍\frac{\mathrm{d}\omega}{(\omega^2-\Omega^2)}. \tag{40}$$

If we arbitrarily fix an upper limit of integration, $\omega_m \gg \Omega$, the second integral, which vanishes as $\omega_m \to \infty$, is very small and may be neglected in comparison with the first, so that we write

$$m_0(\omega_0^2 - \Omega^2) = (e^2/3\pi^2\varepsilon_0 c^3)(\tfrac{1}{3}\omega_m^3 + \Omega^3\omega_m),$$

or

$$(m_0 + e^2\omega_m/3\pi^2\varepsilon_0 c^3)\Omega^2 = \mu_0 - e^2\omega_m^3/9\pi^2\varepsilon_0 c^3. \tag{41}$$

Interaction between the charge and its field may therefore be considered to have two separate effects – an increase in the mass of the oscillating charge and a decrease in its restoring force. The latter is the stronger effect, diverging as ω_m^3, and clearly represents an inherent instability of a charged particle in classical electromagnetic theory. We have already noted this unpleasant phenomenon in chapter 7 and have nothing to add at this point, except to remark that it is not readily eliminated from either the classical or the quantal treatment. The achievements of quantum electrodynamics have taught us that it is possible to live in the shadow of infinities such as this, and to devise systematic procedures for getting the right answer to a variety of problems by pretending they are not there.[4]

I.200

The weaker divergence, on the left-hand side of (41), can be controlled, as the stronger cannot, by assigning non-vanishing radius to the oscillating charge. Suppose, for instance, that the charge is uniformly distributed on the surface of a sphere of radius a. Then it is easy to show that a field varying as $\mathscr{E}_0 \mathrm{e}^{ikx}$ acts on the charge with a force that is not $e\mathscr{E}_0$ but $e\mathscr{E}_0 \sin(ka)/ka$. In the light of this, we may replace the relevant part of (40), which is $\int \Omega^2\, \mathrm{d}\omega$, by $c\int_0^\infty \Omega^2 \sin ka\, \mathrm{d}k/a$, since $\omega = ck$; and in (41) the extra mass is altered by replacing ω_m by $\pi c/2a$. The effective mass is thus $m_0 + e^2/6\pi\varepsilon_0 c^2 a$. Now the electrostatic field energy outside a charged sphere is $e^2/8\pi\varepsilon_0 a$, and if we apply Einstein's mass–energy relation we expect this energy to have a mass $e^2/8\pi\varepsilon_0 c^2 a$, $\frac{3}{4}$ of the amount we have just derived. The history of electrodynamics from about 1880 till Einstein's work (1905) is full of attempts to resolve the precise relationship between electrostatic energy and mass, and is littered with numerical factors of the order of unity. In view of the fact that we have agreed to ignore the gross divergence on the right-hand side of (41), it would have been astonishing for the right answer to have emerged from any subsequent analysis of the problem. Nevertheless the form of the correction to the mass gives confidence in ascribing the mass correction to the field energy, even while we recognize that this is a palliative rather than a cure for the weakness. It is a basic conundrum in particle theory to reconcile the apparent point-character of the electron with its finite mass, and it is here that renormalization theory is particularly successful in sidestepping the issue.

The divergence problem does not arise when an oscillator embedded in a solid interacts with the lattice vibrations (*phonons*) rather than with electromagnetic cavity vibrations (*photons*), since the atomicity of the solid imposes a natural cut-off frequency at the upper end of the vibration spectrum. For a given energy of a vibrational mode, the electric field associated with a lattice vibration is normally very much less than that associated with the same mode in a cavity of the size and shape of the solid, and to this extent the coupling is weaker. The density of states, however, is now $\omega^2/\pi^2 v_s^3$ rather than $\omega^2/\pi^2 c^3$, where v_s is the velocity of sound, perhaps 10^4–10^5 times less than c, and this vastly greater density of states commonly makes phonons dominate photons as the mechanism of energy exchange. By the same token, the enhancement of mass by lattice polarization may be very significant.[(5)] In this case there is no need to invoke Einstein's mass–energy relation – the charged particle elastically deforms the lattice around itself, and enhances the local density; when the particle moves, the excess mass must also move with it. This at any rate is true for slow vibrations of the particle, or slow motion of a charged particle through the solid, to which an expression analogous to (41) applies. But if the particle is caused to vibrate at a frequency much greater than ω_m the neglected last term in (40) is important; indeed, when $\Omega \gg \omega_m$ it cancels the mass-enhancement. Obviously, if one forces the particle to vibrate at a frequency much higher than any possible lattice frequency, the lattice will not follow the motion and there will be no contribution from it to the mass of the particle. An alternative way of visualizing the process is in terms of the response of the particle to an impulsive force. At the instant of application of the impulse, P, the particle jumps forward with velocity P/m_0, responding as though it were a free particle, as already noted in chapter 13. It momentarily leaves behind its clothing of virtual phonons (i.e. the local lattice distortion), but after a time of the order of $1/\omega_m$ these begin to respond and as the lattice reacts in this way it drags the particle back so that eventually particle and lattice distortion proceed together at a lower speed, $P/(m_0 + m_{ph})$, where m_{ph} is the phonon mass-enhancement.

This discussion has perhaps deflected attention from the principal point, which is the description of radiative processes and the meaning of the resulting line broadening. The phenomenon of mass-enhancement can be treated as a quite separate issue, associated especially with the high-frequency end of the vibration spectrum, which is at least partially responsible for whatever mass the oscillating particle happens to exhibit. The line broadening, on the other hand, is something which concerns only those cavity modes whose frequencies lie very close to the oscillator frequency, and there is no reason to doubt the treatment of this process because of flaws in the treatment of other, essentially disconnected, processes.

17 The equivalent classical oscillator

This chapter may be regarded as a prologue to the next or as a brief gathering together of some threads from chapter 13, where it was shown that, under certain not too restrictive conditions, a quantum system responds to an applied force in a way that can be modelled by a set of classical harmonic oscillators, one for every possible transition between energy levels. The impulse-response function, $h(t)$, derived in (13.37), applies either to a system which, before the impulse, is known to be in a stationary state, or to an assembly of identical systems whose previous history has rendered them chaotic in phase. In either case the behaviour of $\langle x(t)\rangle$, considered either as the response of a simple system in the nth quantum state or that part of the average for the response of the assembly that is attributable to the occupation of the nth state, takes the form:

15

$$h_n(t)=\sum_l (f_{ln}/m\omega_{ln})\sin\omega_{ln}t, \tag{1}$$

in which

$$f_{ln}/m\omega_{ln}=2|\langle l|x|n\rangle|^2/\hbar=2|M_{ln}^{(1)}|^2/\hbar, \tag{2}$$

and $\omega_{ln}=(E_l-E_n)/\hbar$. A set of classical oscillators, of which a typical member has mass m/f_{ln} and natural frequency ω_{ln}, would give in sum the same response.

The equivalence of the quantum system and the set of classical oscillators holds when a time-varying force is applied, provided the response is linear, to allow the force to be resolved into a sequence of impulses, each inducing a response of the form (1). One must not apply the argument if the perturbing force is so strong that the occupation of the nth quantum state is significantly altered or (what amounts to much the same) if the assumption of random phase is caused to fail, for $h_n(t)$ will thereby be changed. To estimate how serious this restriction is let us consider a single transition, between the nth and the lth levels, when an electron is perturbed by an electric field $\mathscr{E}\cos\omega t$ and ω is close to ω_{ln}. We put $a_n=1$ and $a_l=0$, and use (13.39) to find how a_l develops:

$$\dot a_l=(ie\mathscr{E}M_{ln}^{(1)}/\hbar)\,e^{-i\omega_{ln}t}\cos\omega t\sim(\tfrac{1}{2}i\mathscr{E}p/\hbar)\,e^{i\delta t}, \tag{3}$$

in which $\delta=\omega-\omega_{ln}$ and $p=eM_{ln}^{(1)}$; a rapidly oscillatory term is neglected. The resulting amplitude of oscillation of a_l at the beat frequency δ is

$\frac{1}{2}\mathscr{E}p/\hbar\delta$, and the criterion for the continued application of (1), which is the criterion of linear response, is simply that a_l must never approach unity:

$$\mathscr{E}p \ll \hbar\delta. \quad (4)$$

Now for an atomic system p is typically the dipole moment resulting from shifting an electron through a distance comparable to the atomic radius, say 3.3×10^{-30} Cm (1 Debye unit). With gross mistuning, e.g. δ equal to the frequency of the line ($\sim 4\times 10^{15}\,\mathrm{s}^{-1}$ for an optical transition) $\mathscr{E}$ must approach $10^{11}\,\mathrm{V\,m}^{-1}$ to initiate non-linear processes. This is an enormous field strength, such as would be found in a beam of radiation whose power flux was about $10^{19}\,\mathrm{W\,m}^{-2}$. Huge as this is, it is not beyond the flux obtainable in short pulses from the largest lasers, and at this power level virtually every physical system will respond non-linearly, even if it is not immediately destroyed. With much less power and better tuning non-linear effects are relatively easy to obtain; thus with δ 10^5 times smaller than ω_{ln}, the required flux is only $10^8\,\mathrm{W\,m}^{-2}$ – out of the question with monochromatic radiation before the invention of lasers, but readily available since. If there were no radiative loss, and the spectral line were consequently perfectly sharp, (4) would hold for all δ, however small; under exact tuning conditions any value of $\mathscr{E}$ would suffice to produce non-linear effects in the response. Such a state of affairs is not very different from the experimental conditions prevailing in nuclear magnetic resonance, where very low power levels can cause saturation of the line. In spectroscopy at optical frequencies, however, (4) begins to fail when δ is comparable to the line width and quite sizeable values of $\mathscr{E}$ are allowable. Loss of energy by radiative transitions then limits the build-up of a_l and encourages linear response.

180

The orders of magnitude just arrived at explain the interest, since the invention of lasers, in the theory of non-linear optical processes, which we shall not consider here;[1] but they also show that for many purposes the assumption of linearity is fully justified, and with it the use of the equivalent classical oscillator as an aid to description of processes and for problem-solving. Within the linear assumption the changes in the occupation probabilities $a_n a_n^*$ are so small that one need not worry about transitions involving three or more levels, such as the climbing of the ladder of oscillator levels discussed in chapter 13. Indeed the harmonic oscillator is an especially sensitive system from this point of view on account of the even spacing of the levels. More usually the significant excitation of a system from the nth level by radiation closely tuned to ω_{nl} will find it in the lth level with nowhere else to go except back again, all other potentially accessible levels being at the wrong spacing for near-resonance. Systems with equally spaced levels, like the infinite set for a harmonic oscillator, or the $2s+1$ levels for a particle of spin s, respond in a way which has marked classical analogies, even when strongly excited, and thereby are not inaccessible to exact theory. With other systems non-linearity only begins to involve more than two levels with a much stronger perturbing force, such that (4) fails for a whole chain of connected levels. The theoretical problem then begins to be formidable except to the sledgehammer approach of a computer. The intermediate case, however, can be approximated by considering only the two levels which are near resonance with the perturbing force, and this provides the incentive for detailed analysis of the two-level system in the next chapter. To take things in proper order, however, let us first develop some consequences of the equivalent classical oscillator theory.

The f-sum rule

15

It was noted in chapter 13 that when an impulse P is applied to a quantum system the immediate response may be derived by neglecting V in the time-dependent Schrödinger equation and consequently treating the particle as free. Moreover it is easily shown that the result (13.30) implies that the expectation value of the momentum is increased by P, while the position is unchanged, just as for a classical particle. Now, according to (1), when unit impulse is applied to the system in its nth state, the instantaneous response of the velocity may be written

$$\dot{\hbar}_n(0) = \sum_l f_{ln}/m,$$

and it is this that must match the response of a classical particle, which is $1/m$. It follows that when a real system, in which only one particle is affected by the applied force, is replaced by a set of equivalent oscillators, each acted upon by the same force, the oscillator strength, f_{ln}, involved in transitions from a given level, n, to any other level, l, must obey the f-sum rule:

$$\sum_l f_{ln} = 1. \qquad (5)$$

We have already had an example in the harmonic oscillator, where a uniform force can induce transitions from n only to $n \pm 1$, and where $f_{n+1,n} = n+1$ and $f_{n-1,n} = -n$, so that (5) is obviously true. Let us explicitly verify the rule in another case which is easy to treat, a particle confined to a limited stretch of a line, $0 < x < a$, in which range $V = 0$. Then the nth wave-function (normalized) is $(2/a)^{\frac{1}{2}} \sin(n\pi x/a)$ and its energy is $n^2(\pi^2\hbar^2/2ma^2)$. Consequently $\omega_{ln} = \pi^2\hbar(l^2 - n^2)/2ma^2$ and

$$\begin{aligned} \langle l|x|n\rangle &= (2/a)\int_0^a x \sin(l\pi x/a)\sin(n\pi x/a)\,\mathrm{d}x \\ &= -8nla/[\pi^2(l^2-n^2)^2] \text{ if } l-n \text{ is odd}, \qquad (6) \\ &= 0 \text{ if } l-n \text{ is even.} \end{aligned}$$

Hence $f_{ln} = 64n^2l^2/[\pi^2(l^2-n^2)^3]$ if $l-n$ is odd. Let us evaluate f_{ln}, for transitions from the state $n = 4$:

$l =$	1	3	5	7	9	11	13	15
$f_{l4} =$	−0.0307	−2.7234	3.5581	0.1415	0.0306	0.0108	0.0049	0.0026

The sum of the f_{l4} up to this point is 0.9954 and the higher terms account for the missing fraction. If we had considered transitions from any other state, the f-values would of course have all been different, but the sum is always unity.

It will be noted that, as with the harmonic oscillator, some of the terms are greater than unity, indicating a stronger response than a classical oscillator could provide, and that downward transitions have negative strength. The contribution of these to the impulse response function is opposite in sign to that of a classical bound particle having positive mass. There are consequences which will play an important role in what follows. For example, upward and downward transitions contribute oppositely to the polarization of a molecular system by a steady field, upward transitions giving rise to a positive electrical susceptibility, downward to a negative. And when the response function exhibits decay as a result of radiative transitions or collisions, the upward transitions reveal this as dissipation, but the downward as negative dissipation, or gain. The action of masers and lasers depends on this sign reversal.

The f-sum rule expressed in (5) applies to a single particle, but a generalization is straightforward. We consider the commonest case, that of electric dipole transitions, where the force is applied by a uniform electric field, $\mathscr{E}$; different particles in a complex system experience different forces, $q\mathscr{E}$, according to their charge q. A free particle of charge q_i and mass m_i, acted on by an impulsive field, $P_e = \int \mathscr{E}\,\mathrm{d}t$, responds with an instantaneous velocity change $P_e q_i/m_i$, and an instantaneous change in electric current $P_e q_i^2/m_i$. If the response of the nth state of the system is to be simulated by a set of classical oscillators, all carrying charge e but with masses m/f_{ln}, then

$$\sum_l f_{ln} e^2/m = \sum_i q_i^2/m_i,$$

i.e.
$$\sum_l f_{ln} = \sum_i (q_i/e)^2/(m_i/m). \tag{7}$$

For a system composed of N identical particles, electrons for instance, it is natural to choose e and m to be the electronic charge and mass, whereupon (7) takes the form $\sum_l f_{ln} = N$.

It often happens that in the system to which the f-sum rule is applied some of the particles are more or less free, such as conduction electrons in a metal, some are loosely bound (electrons near the top of the valence band in a semiconductor), some tightly bound (inner shell electrons) and some very tightly bound indeed by these standards (individual nucleons in the nuclei). The corresponding oscillator frequencies cover a huge frequency range, but if they fall into well-separated bands it may be possible to neglect the high-frequency oscillators when considering low-frequency phenomena; nuclear structure is of small consequence to the conduction properties of solids, and indeed the nucleus and the inner electron shells are often treated as a single structureless charged particle, the ion core. There is, however, no blanket rule to cover neglect of tightly bound structures; the polarization of an atom or an insulating solid may be dominated by the equivalent oscillators of low frequency, but inner shells often make a far from negligible contribution. Care is always needed when the oscillator spectrum is truncated.

Static polarizability

A typical equivalent oscillator, of strength f_{ln}, and characteristic frequency ω_{ln}, must be assigned a restoring force constant μ_{ln} equal to $m\omega_{ln}^2/f_{ln}$. Under the influence of a steady field $\mathscr{E}$ the displacement is $e\mathscr{E}/\mu_{ln}$ and the resulting dipole moment $e^2\mathscr{E}/\mu_{ln}$. Hence the total dipole moment acquired by the system in its nth state may be written

$$p = (e^2\mathscr{E}/m)\sum_l f_{ln}/\omega_{ln}^2. \tag{8}$$

If the system considered is a single atom or molecule the molecular polarizability α is defined by the equation $p = \alpha\varepsilon_0\mathscr{E}$. With N molecules per unit volume the polarization per unit volume $P = N\alpha\varepsilon_0\mathscr{E} = \kappa\varepsilon_0\mathscr{E}$, where κ is the volume susceptibility, $N\alpha$†. From (8),

$$\alpha_n = (e^2/m)\sum_l f_{ln}/\omega_{ln}^2. \tag{9}$$

The example already studied, the particle moving on a limited stretch of line, serves to illustrate this result, and we shall compute the polarizability in the ground state, $n = 1$. From (6), $f_{l1} = 64l^2/[\pi^2(l^2-1)^3]$ for even l, and $\omega_{l1} = (l^2-1)(\pi^2\hbar^2/2ma^2)$; hence

$$\begin{aligned}\alpha_0 &= \frac{256me^2a^4}{\pi^6\hbar^2}\sum_{l\,\text{even}} l^2/(l^2-1)^5 \\ &= \frac{me^2a^4}{\pi^4\hbar^2}(42\,696\,663 + 54\,652 + 1778 + 167 + 27 + 6 + 2 + 1) \\ &\quad \times 10^{-8} = 0.42753296\, me^2a^4/\pi^4\hbar^2.\end{aligned} \tag{10}$$

To calculate α_0 directly Schrödinger's equation must be solved for the potential $V = -\mathscr{E}x$, but the wave-function need be correct only to first order in $\mathscr{E}$. It may be checked by substitution that the following solution is adequate:

$$\psi = (1-\beta x)\sin kx + \beta kx^2\cos kx, \tag{11}$$

where $\hbar^2k^2/2m = E$ and $\beta = e\mathscr{E}ma^2/2\pi^2\hbar^2$. The zeros of ψ are not exactly

† Strictly $\boldsymbol{\mathscr{E}}$ and $\boldsymbol{P}$ are vectors and κ is a second-rank tensor. In the three-dimensional case one must recognize that the matrix element $\langle l|x|n\rangle$ must be treated as a vector $\langle l|x_i|n\rangle$ and that f_{ln} is a second-rank tensor, $(2m\omega_{ln}/\hbar)\langle l|x_i|n\rangle\langle l|x_j|n\rangle^*$, conferring tensorial anisotropy on the reciprocal mass $1/m_{ln} = f_{ln}/m$. This is a complication which must be taken account of in many real computations, but does not affect the physical principles discussed here.

at 0 and a, but at 0 and $X(=a(1-\beta a))$; this does not matter, however, provided the same stretch of line is used to calculate $\langle x\rangle$ before and after $\mathscr{E}$ is applied, the difference yielding α correct to first order in $\mathscr{E}$. Now, from (11), $\int_0^X \psi^2\,dx = \frac{1}{2}a(1-2\beta a)$ and $\int_0^X x\psi^2\,dx = \frac{1}{4}a^2[1-(10/3-5/\pi^2)\beta a]$. It follows that after $\mathscr{E}$ is applied

$$\langle x\rangle = \int_0^X x\psi^2\,dx \Big/ \int_0^X \psi^2\,dx = \tfrac{1}{2}a[1-(4/3-5/\pi^2)\beta a],$$

while when $\mathscr{E}=0$, $\langle x\rangle = \frac{1}{2}X$. The shift is $\frac{1}{2}\beta a^2(5/\pi^2-1/3)$, and hence

$$\alpha = (me^2a^4/\pi^4\hbar^2)(5/4-\pi^2/12) = (me^2a^4/\pi^4\hbar^2)\times 0.42753297,$$

in agreement with (10).

It is tempting to conclude from the appearance of (10) that the transition to the next level so dominates α for all others to be negligible unless a very precise value is needed. And the temptation is compounded when we apply this principle to helium, for which the resonance line at a wavelength of 59 nm corresponds to the excitation of one electron only from $1s$ to $2p$ (by symmetry the excitation $1s-2s$ has an f-value of zero). If we assume that this provides the only significant equivalent oscillator, with f near unity, the resulting estimate of the volume susceptibility κ at NTP is 8.4×10^{-5}, comparing very favourably with the measured value of 7.0×10^{-5}. The same argument applied to xenon, however, whose resonance line is at 124 nm, would suggest that κ should be about 4 times as great as in helium, whereas in fact it is 20 times. Obviously the large number of electrons in the atom leads to a much higher density of excited states which, even if contributing little individually, add up to an overwhelming total.

Another warning is provided by calculating the f-values for the transitions from the ground state of hydrogen, $1s$, to the higher p-states, $2p$, $3p$ etc., which are the only transitions for which f does not vanish. In this case the wave-functions are well known and the integrations straightforward, if tedious.[2] It is found that $f_{21}=0.4162$, $f_{31}=0.0791$, $f_{41}=0.0290\ldots$, with a sum $\Sigma_{n=2}^{\infty} f_{n1}=0.5641$. The f-sum is not unity because we have not taken into account the continuum of unbound states with positive energy, and it is clear that they contribute nearly half the oscillator strength. These examples serve to emphasize once more how cautious one must be in truncating the oscillator set. The lowest lying states may allow the order of magnitude of α to be estimated, but for anything approaching exactitude there are no short cuts.

It will be appreciated that in the calculation of the static polarizability there is no preferential weighting of certain equivalent oscillators by exciting them near resonance, except in so far as the presence of ω_{ln}^2 in the denominator of (9) makes low-energy transitions particularly important. As soon as we concern ourselves with resonant responses, however, we may narrow our vision, as already discussed, to exclude all but those levels to which there is an enhanced transition probability.

18 The two-level system

We now specialize the results of the last chapter, to consider in some detail the behaviour of a system in which only two levels are significantly occupied. In proton spin resonance, for example, all the translational states of the proton are irrelevant under ideal conditions since they are not affected by the oscillating magnetic field that induces transitions between the spin states; and in the case of a particle tunnelling between two wells the lowest two states are much closer in energy than they are to other excited states, so that at the low frequencies required to change the occupations of these lowest states the others remain uninvolved. In neglecting all others we do not imply that they do not exist – we do not, for example, assign an oscillator strength of unity to the transition between the states of interest – we simply ignore them and define the state of the system by the two complex amplitudes, a_1 and a_2, of the eigenfunctions, $\chi_1(\boldsymbol{r})$ and $\chi_2(\boldsymbol{r})$, taken as real. At any instant the wave-function takes the form

$$\Psi(\boldsymbol{r}, t) = a_1\chi_1\,\mathrm{e}^{-\mathrm{i}E_1 t/\hbar} + a_2\chi_2\,\mathrm{e}^{-\mathrm{i}E_2 t/\hbar}, \tag{1}$$

in which $E_2 > E_1$. If $E_2 - E_1 = 2\Delta_0$ the resonance frequency is $2\Delta_0/\hbar$. Under the influence of disturbing forces a_1 and a_2 change with time, but if the wave-function is initially normalized $a_1a_1^* + a_2a_2^* = 1$ always.

A convenient model for developing the argument is a one-dimensional two-well system for which χ_1 and χ_2 are functions of y only. We shall find that most of the results are expressible in terms of three characteristic dipole moments:

$$\begin{aligned} p_1 &= \langle 1|ey|1\rangle \\ p_2 &= \langle 2|ey|2\rangle \\ p_{12} &= \langle 2|ey|1\rangle \end{aligned} \tag{2}$$

By using these, the results will appear to depend very little on the details of the model, as indeed is the case; we use a particular two-well system to make the derivations explicit and elementary, but shall not hesitate to apply the results in a wider context, even to spin resonance. It should be noted, however, that for all its practical importance we shall have little to say on that topic, other than to give some justification for the classical treatment in chapter 8.

I.218

General theory

When a time-varying electric field, $\mathscr{E}(t)$, is applied parallel to y, a_1 and a_2 change in accordance with (13.40):

$$\left.\begin{aligned} \dot{a}_1 &= (\mathrm{i}\mathscr{E}/\hbar)(p_1 a_1 + p_{12} a_2\, \mathrm{e}^{-2\mathrm{i}\Delta_0 t/\hbar}) \\ \text{and} \qquad \dot{a}_2 &= (\mathrm{i}\mathscr{E}/\hbar)(p_2 a_2 + p_{12} a_1\, \mathrm{e}^{2\mathrm{i}\Delta_0 t/\hbar}) \end{aligned}\right\} \tag{3}$$

A field varying as $\mathscr{E}\cos\omega t$ can be written as $\frac{1}{2}\mathscr{E}(\mathrm{e}^{\mathrm{i}\omega t}+\mathrm{e}^{-\mathrm{i}\omega t})$, and each equation then contains terms involving three frequencies, $|\omega|$ and $|\omega \pm 2\Delta_0/\hbar|$. Close to resonance the slowest vibration, $|\omega - 2\Delta_0/\hbar|$, dominates the evolution of a_1 and a_2, and the others only superimpose rapid oscillations of small amplitude. If these are ignored the resulting simplified versions of (3) are readily solved. Writing δ for $\hbar\omega/2\Delta_0 - 1$, δ_0 for $p_{12}\mathscr{E}/2\Delta_0$ and τ for $2\Delta_0 t/\hbar$, we have in this approximation:

$$\mathrm{d}a_1/\mathrm{d}\tau = \tfrac{1}{2}\mathrm{i}\delta_0 a_2\, \mathrm{e}^{\mathrm{i}\delta t} \quad \text{and} \quad \mathrm{d}a_2/\mathrm{d}\tau = \tfrac{1}{2}\mathrm{i}\delta_0 a_1\, \mathrm{e}^{-\mathrm{i}\delta\tau}. \tag{4}$$

On substituting the trial solution

$$a_1 = c_1\, \mathrm{e}^{\mathrm{i}\gamma\tau}, \qquad a_2 = c_2\, \mathrm{e}^{\mathrm{i}(\gamma-\delta)\tau}, \tag{5}$$

it is found that

$$\left.\begin{aligned} \gamma(\gamma-\delta) &= \tfrac{1}{4}\delta_0^2, \quad \text{or } \gamma = \tfrac{1}{2}[\delta \pm (\delta^2+\delta_0^2)^{\frac{1}{2}}] \\ \text{and} \qquad c_2/c_1 &= 2\gamma/\delta_0. \end{aligned}\right. \tag{6}$$

The two solutions (5) are conveniently written in the form:

$$\left.\begin{aligned} a_1 &= \mathrm{e}^{\mathrm{i}\gamma_1\tau}, \qquad a_2 = (2\gamma_1/\delta_0)\, \mathrm{e}^{\mathrm{i}\gamma_2\tau} \\ \text{and} \qquad a_1 &= \mathrm{e}^{-\mathrm{i}\gamma_2\tau}, \qquad a_2 = -(2\gamma_2/\delta_0)\, \mathrm{e}^{-\mathrm{i}\gamma_1\tau}, \end{aligned}\right\} \tag{7}$$

in which $\gamma_1 = \frac{1}{2}[(\delta^2+\delta_0^2)^{\frac{1}{2}}+\delta]$ and $\gamma_2 = \frac{1}{2}[(\delta^2+\delta_0^2)^{\frac{1}{2}}-\delta]$. The general solution is a linear combination:

$$\left.\begin{aligned} a_1 &= C_1\, \mathrm{e}^{\mathrm{i}\gamma_1\tau} + C_2\, \mathrm{e}^{-\mathrm{i}(\gamma_2\tau+\phi)} \\ \text{and} \qquad a_2 &= (2/\delta_0)[C_1\gamma_1\, \mathrm{e}^{\mathrm{i}\gamma_2\tau} - C_2\gamma_2\, \mathrm{e}^{-\mathrm{i}(\gamma_1\tau+\phi)}]. \end{aligned}\right\} \tag{8}$$

Since the absolute phases of a_1 and a_2 are irrelevant C_1 and C_2 may be taken as real, and ϕ as an arbitrary phase. In the normalized solution $a_1a_1^* + a_2a_2^* = 1$, and hence

$$\left.\begin{aligned} \gamma_1 C_1^2 + \gamma_2 C_2^2 &= \delta_0^2/4\delta', \\ \text{where} \qquad \delta' = \gamma_1 + \gamma_2 &= (\delta^2+\delta_0^2)^{\frac{1}{2}}. \end{aligned}\right. \tag{9}$$

The dipole moment associated with the general solution is readily evaluated:

$$\langle p\rangle = e\langle y\rangle = a_1a_1^*p_1 + a_2a_2^*p_2 + (a_1a_2^*\, \mathrm{e}^{\mathrm{i}\tau} + a_2a_1^*\, \mathrm{e}^{-\mathrm{i}\tau})p_{12}. \tag{10}$$

The first two terms are non-oscillatory except as a result of changes in a_1 and a_2 caused by external influence, and they are present only if the stationary states themselves have permanent dipole moments. They are absent in proton resonance but present when a particle tunnels between two unequal potential wells. The latter case will arise later in the chapter, but for the moment we shall concentrate on the oscillatory dipole resulting from superposition of the states. By using (9), the last term of (10) may be evaluated for the general solution (8):

$$\langle p\rangle/p_{12} = \mathrm{Re}\,[\mathrm{e}^{\mathrm{i}(1+\delta)\tau}\{[(\delta_0/\delta')^2-(4C_1C_2)^2]^{\frac{1}{2}} + (4C_1C_2/\delta_0)(\gamma_1\,\mathrm{e}^{-\mathrm{i}(\delta'\tau+\phi)}-\gamma_2\,\mathrm{e}^{\mathrm{i}(\delta'\tau+\phi)})\}]. \quad (11)$$

This result has the simple geometrical interpretation shown in fig. 1. The quantity in braces, { }, represents on the complex plane an elliptical trajectory which is the projection of uniform motion round a tilted circle; the

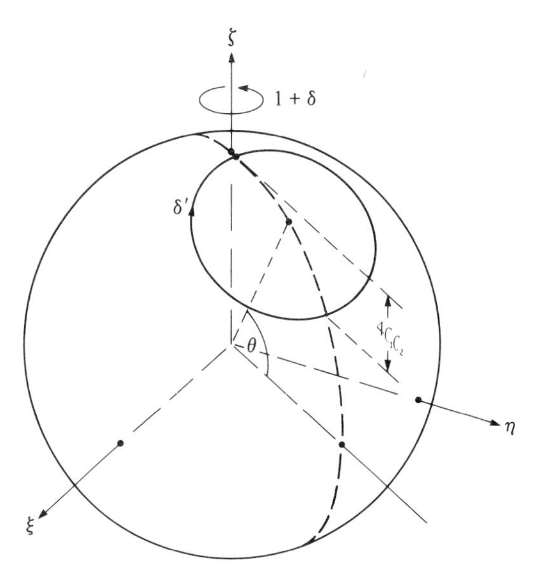

Fig. 1. Geometrical representation of (11) as the projection, onto the horizontal (ξ, η) plane, of motion round a circular orbit on the surface of a sphere.

radius and tilt of the circle can be inferred from the major and minor semi-axes $(4C_1C_2/\delta_0)(\gamma_1\pm\gamma_2)$, while the centre of the ellipse is at $[(\delta_0/\delta')^2-(4C_1C_2)^2]^{\frac{1}{2}}$. It is easily shown that the circle is a plane section of a sphere of unit radius, at an angle $\theta=\tan^{-1}(\delta/\delta_0)$, and that the vertical excursion is $4C_1C_2$. The representative point moves relatively slowly round this circle, with angular velocity δ', while the whole diagram spins at the driving frequency, $1+\delta$. The ξ-co-ordinate of the point is $\langle p\rangle/p_{12}$, and it executes a harmonic oscillation at something like the beat frequency $(E_2-E_1)/\hbar$ with an amplitude which is modulated at something like the detuning frequency, δ, or more precisely δ'. This behaviour is exactly the same as that described in chapter 8 and illustrated in fig. 8.7 – the motion of a gyromagnetic top in a steady vertical magnetic field, when perturbed by a weak field oscillating or rotating in the horizontal plane. It is the exactness

I.220

of the analogy which allows us, just as with the harmonic oscillator, to discuss in classical pictorial terms what would seem to be the strictly quantal phenomenon of spin resonance.

The diagrammatic representation of fig. 1 acquires new significance when described in terms of Cartesian co-ordinates. At any instant let Ψ be $a_1\chi_1+a_2\chi_2$, the phase factors $e^{-iE_{1,2}t/\hbar}$ being incorporated in a_1 and a_2, and let this wave-function be represented on the unit sphere by a point defined as follows:

$$\zeta = a_2a_2^* - a_1a_1^*, \qquad \rho = \xi + i\eta = 2a_1a_2^*. \tag{12}$$

Thus the upper pole of the sphere ($\zeta = 1$) represents the pure excited state, the lower ($\zeta = -1$) the pure ground state, while the azimuthal angle of any other point defines the instantaneous phase difference between the two components; the wave-function is automatically normalized, $a_1a_1^* + a_2a_2^* = 1$. Now from the last term in (10) it is clear, since $e^{i\tau}$ is incorporated in $a_1a_2^*$, that ξ so defined represents $\langle p\rangle/p_{12}$, and that the evolution of a_1 and a_2 causes the point to move in the manner required by (11). What is rather special about spin resonance, as compared with the double-well system that is also described by a simple two-component wave-function, is that all the Cartesian components are meaningful, and not just the ξ-component. In fact we have done no more than take over from the quantum theory of spin a pictorial representation that serves a much wider range of interest. The Pauli matrices:[1]

$$\sigma_\xi = \frac{\hbar}{2}\begin{pmatrix}0 & 1\\ 1 & 0\end{pmatrix}, \qquad \sigma_\eta = \frac{\hbar}{2}\begin{pmatrix}0 & -i\\ i & 0\end{pmatrix}, \qquad \sigma_\zeta = \frac{\hbar}{2}\begin{pmatrix}1 & 0\\ 0 & -1\end{pmatrix},$$

generate expectation values for the three components of angular momentum of the same form as (12):

$$\langle L_\xi\rangle = \frac{\hbar}{2}(a_1^* \quad a_2^*)\begin{pmatrix}0 & 1\\ 1 & 0\end{pmatrix}\begin{pmatrix}a_1\\ a_2\end{pmatrix} = \frac{\hbar}{2}(a_1^*a_2 + a_1a_2^*)$$

$$\langle L_\eta\rangle = \frac{\hbar}{2}(a_1^* \quad a_2^*)\begin{pmatrix}0 & -i\\ i & 0\end{pmatrix}\begin{pmatrix}a_1\\ a_2\end{pmatrix} = -\frac{i\hbar}{2}(a_1^*a_2 - a_1a_2^*)$$

$$\langle L_\zeta\rangle = \frac{\hbar}{2}(a_1^* \quad a_2^*)\begin{pmatrix}1 & 0\\ 0 & -1\end{pmatrix}\begin{pmatrix}a_1\\ a_2\end{pmatrix} = \frac{\hbar}{2}(a_1a_1^* - a_2a_2^*)$$

In the light of this there is little need to justify in greater detail the use of classical arguments in chapter 8 to describe proton resonance, and we shall continue to rely on their validity on those occasions when the discussion of that chapter needs extension. We shall not consider particles of spin other than $\frac{1}{2}$, which have more than two equally spaced levels and which bridge the gap between spin-$\frac{1}{2}$ particles (two levels) and harmonic oscillators (unlimited number of levels).

The double-well model

It is now convenient to keep in mind a model system, having two closely spaced levels, which is closer in form than is the spin-$\frac{1}{2}$ particle to the systems with which most of this and the next chapter will be concerned. We consider first the system shown in fig. 2(*a*) in which the one-dimensional

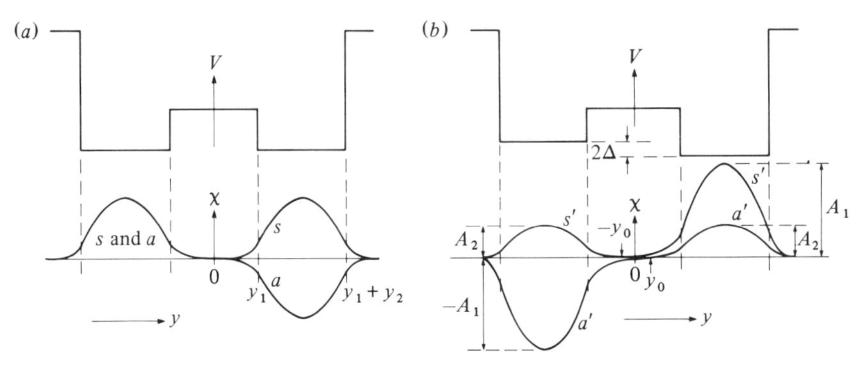

Fig. 2. The double-well model: (*a*) equal wells (above) giving rise to symmetrical and antisymmetrical wavefunctions (below); (*b*) unequal wells for which the lower state (s′) has larger amplitude in the deeper well, and the upper state (a′) larger amplitude in the shallower well.

potential wells are equal in depth and width. If the barrier is high and wide enough to preclude tunnelling, the wells are uncoupled and the ground state wave-function may have arbitrary amplitudes in each of them, but once they are coupled by the possibility of tunnelling there are two low-lying stationary states, one symmetrical (even) and the other antisymmetrical (odd), the former having lower energy than the latter. Inside the wells the wave-function is a sinusoid, at the outside edges it is exponential, and in the barrier it takes the form $\cosh \mu y$ in the even case, $\sinh \mu y$ in the odd, μ being determined by the height of the barrier. Because of the spread of χ beyond the sides of the well, the half-wavelength of the sinusoid in the well is slightly more than the width y_2. It is a very good approximation to take this small excess to be the sum of the values of χ/χ' at the edges, i.e. to extrapolate the sinusoid linearly to zero and take the effective edge to lie where χ hits the axis. At the outside edges the excess is constant and can be neglected, but the variations of the excess on the inside edges, depending on the tunnelling probability, are responsible for the shift of energy levels, which is the essential feature of the model. Thus if $\chi = \cosh \mu y$ in the barrier, χ/χ' at the barrier edge ($y = y_1$) takes the value $(\coth \mu y_1)/\mu$, rather than the value $1/\mu$ which it would take if y_1 were so large as to make the barrier impenetrable. Tunnelling therefore increases the effective width by $(\coth \mu y_1 - 1)/\mu$ and correspondingly lowers the kinetic energy of the even state, which is $\hbar^2 k^2/2m$ for a sinusoidal wave-function of wavenumber k. Hence

$$\Delta E/E = 2\Delta k/k = -2(\coth \mu y_1 - 1)/\mu y_2,$$

y_2 being the width of the well. Conversely, in the odd state the effective width is narrowed, $\sinh \mu y$ being steeper than $e^{\mu y}$, and the energy is raised by nearly the same amount. The levels are thus at $\pm\Delta_0$ relative to the uncoupled wells, where

$$\left.\begin{array}{ll} & \Delta_0 = C(\coth \mu y_1 - 1) \\ \text{or} & C(1 - \tanh \mu y_1) \end{array}\right\} \sim 2C\,\mathrm{e}^{-2\mu y_1} \tag{13}$$

and
$$C = \pi^2\hbar^2/\mu m y_2^3.$$

When the system is set up in a mixture of the two states, the beating of the wave-functions results in a sinusoidal exchange of the particle between the wells at a frequency of $2\Delta_0/\hbar$, i.e. $(\pi^2\hbar/\mu m y_2^3)\,\mathrm{e}^{-2\mu y_1}$. If one finds a classical picture helpful, the pre-exponential factor is, apart from a constant of order unity, the frequency at which a particle of the same energy strikes the inner edge of its cell, and the exponential represents its chance of penetrating the barrier at each attempt (strictly, the amplitude reduction across the barrier). But this is little more than a mnemonic.

When the potential wells are unequal, as in fig. 2(*b*), the centre of the wave-function in the barrier is shifted from the origin to some new position y_0. It is now convenient to synthesize a solution and then to determine what problem has been solved. For example, let the even solution be modified so that the wave-function in the barrier is $\cosh \mu(y + y_0)$, the minimum being moved from the centre to $-y_0$. The same approximation as leads to (13) gives a different kinetic energy shift in the two wells, $-2C\,\mathrm{e}^{-2\mu(y_1 \pm y_0)}$ or $-\Delta_0\,\mathrm{e}^{\pm 2\mu y_0}$, the upper sign referring to the left-hand well in which the kinetic energy is lower than in the right-hand well. The difference in kinetic energy must be compensated by the difference in potential energy, 2Δ, from which it follows that this solution obtains when $2\Delta = \Delta_0(\mathrm{e}^{2\mu y_0} - \mathrm{e}^{-2\mu y_0})$. Hence

$$\Delta = \Delta_0 \sinh(2\mu y_0). \tag{14}$$

The overall lowering of the energy is obtained by summing kinetic and potential contributions in either well; thus from the right-hand well the energy is found to be $-E$, where

$$E = \Delta + \Delta_0\,\mathrm{e}^{-2\mu y_0} = \Delta_0 \cosh(2\mu y_0) = (\Delta_0^2 + \Delta^2)^{\frac{1}{2}}. \tag{15}$$

In the upper state, for which the wave-function in the barrier is $\sinh \mu(y + y_0)$, the energy is $+E$.

The wave-function in the wells is so nearly sinusoidal that a single number serves to define the amplitude in each well. Since the gradient of the sinusoid as it approaches zero is determined by its amplitude, it follows that the values of χ' at the inner edges of the wells fix the relative amplitudes. If these are A_1 and A_2 in the right and left wells when the system is in its lower state, they are A_2 and $-A_1$ in the upper state, and

$$\left.\begin{array}{ll} & A_1/A_2 = \sinh \mu(y_1 + y_0)/\sinh \mu(y_1 - y_0) \\ \text{or} & \cosh \mu(y_1 + y_0)/\cosh \mu(y_1 - y_0) \end{array}\right\} \sim \mathrm{e}^{2\mu y_0} = \frac{E + \Delta}{\Delta_0} = \frac{\Delta_0}{E - \Delta}. \tag{16}$$

Hence $$A_1^2/A_2^2 = (E+\Delta)/(E-\Delta), \tag{17}$$

or $$A_1 = (1+\Delta/E)^{\frac{1}{2}}/\surd 2 \quad \text{and} \quad A_2 = (1-\Delta/E)^{\frac{1}{2}}/\surd 2, \tag{18}$$

if the amplitudes are normalized so that $A_1^2 + A_2^2 = 1$. We may now relate the dipole moments defined in (2) to a standard moment p_0 obtained by placing the charge centrally in one of the wells. Then

$$\left.\begin{aligned} p_1 = -p_2 = p_0(A_1^2 - A_2^2) = p_0\Delta/E \\ \text{and} \qquad p_{12} = 2p_0 A_1 A_2 = p_0\Delta_0/E. \end{aligned}\right\} \tag{19}$$

For equal wells p_1 and p_2 vanish and $p_{12} = p_0$. When the wells are unequal

$$p_1^2 + p_{12}^2 = p_0^2. \tag{20}$$

The results expressed in (15), (16) and (17) are of the same form as those describing coupled resonant circuits. Thus (15) is to be compared with (12.11) and fig. 12.2, frequencies in the circuits being replaced by energies in the double-well system, while the tunnelling probability as measured by Δ_0 is the analogue of the coupling parameter κ for the circuits. Similarly the ratio of the energy content of the two circuits when a pure normal mode is excited, as given by (12.10), finds an exact parallel in the ratio of probabilities of finding the particle in the two wells, as given by (17). Neither for the circuits nor for the double-well are the results sensitive to the details of the system. The shape of the potential wells does not matter – it is enough that they should take the same form apart from a vertical displacement, and that the coupling should be weak.[(2)] Even the shapes of the wells need not be identical provided care is taken in defining the amplitudes $A_{1,2}$ so that their squares represent probabilities of finding the particle in each well.

I.366

Response functions

Application of a uniform electric field $\mathscr{E}$ superposes a linear variation of potential on the double well. The tilt of the bottom of each well is of secondary importance compared to the shift of the mean levels, and the effect of the field may be adequately described by a change of Δ by $p_0\mathscr{E}$. If the system was initially in its lower state and is able after the application of $\mathscr{E}$ to revert to this state its dipole moment is changed by δp_1, which (19) shows to be $p_0^2\Delta_0^2\mathscr{E}/E^3$. In the upper state the result is the same except for a change of sign. The pure states therefore have polarizabilities

$$\alpha = \mp p_0^2\Delta_0^2/E^3, \tag{21}$$

the upper sign referring to the upper state. When the wells are equal (21) becomes $\mp p_0^2/\Delta_0$, and this is the condition of maximum polarizability.

The arguments of the last chapter find a very simple application here. There is only one f-value of significance and the force constant for the equivalent oscillator is $E/\langle 2|y|1\rangle^2$, i.e. e^2E/p_{12}^2, giving a polarizability of

p_{12}^2/E; (19) shows that this is the same as (21). As the coupling between equal wells is reduced E falls while p_{12} is constant, and α increases without limit. The value of f_{12}, according to (17.2), is $4m\Delta_0 p_{12}^2/\hbar^2 e^2$, roughly equal to the tunnelling factor $e^{-2\mu y_1}$, and therefore very small. Here is an example of how the overwhelmingly most significant equivalent oscillator may account for only a tiny fraction of the oscillator strength; its low frequency and hence its weak force constant are responsible for its importance.

The result expressed by (21) holds only for a pure state. An assembly of identical systems in thermal equilibrium has its mean polarizability $\bar{\alpha}$ determined by the excess of systems in the lower state relative to those in the upper state, for the two states have equal and opposite polarizabilities. When Δ_0 is small the effect of the high individual values of α is largely compensated by the small excess in the lower state, and in fact $\bar{\alpha}$ tends to become independent of Δ_0. Let us suppose that before $\mathscr{E}$ is applied half the systems have an inequality of their wells of one sign, Δ, and half are reversed in sign, $-\Delta$, so that the resultant total dipole moment is zero. Consider the first half only, of which fractions $e^{\mp E/k_BT}/(e^{E/k_BT}+e^{-E/k_BT})$ will be in the two states, giving a mean moment

$$\bar{p}/p_0 = (\Delta/E)\tanh(E/k_BT). \tag{22}$$

The mean polarizability follows by differentiating and setting $d\Delta/d\mathscr{E}$ equal to p_0:

$$\bar{\alpha} = d\bar{p}/d\mathscr{E} = p_0^2\, d(\bar{p}/p_0)/d\Delta = p_0^2[(\Delta_0^2/E^3)\tanh(E/k_BT) + (\Delta^2/E^2k_BT)\,\mathrm{sech}^2(E/k_BT)]. \tag{23}$$

The other half of the assembly gives an identical result, so that (23) is correct for the whole assembly. The two terms in the square brackets reveal two distinct sources of polarization by the field. The first reflects the polarizability of each state (21), reduced by the factor $\tanh(E/k_BT)$ which is the excess fraction of systems in the lower state. The second results from the change in E, and therefore of the Boltzmann factor, when $\mathscr{E}$ is applied. Whereas the dipole moment due to the first term appears virtually instantaneously, since the energy levels adjust themselves immediately, the second depends on systems making transitions between states and thereby approaching a new condition of statistical equilibrium; if the systems are isolated from extraneous disturbances and have to rely on radiative processes, a long time may elapse before the second term makes itself felt.

We have, however, oversimplified in suggesting that the first process is instantaneous, for it is clear that the sudden application of a weak field will not immediately change the occupation of the wells, and the total dipole moment must therefore suffer no sudden change. When $\mathscr{E}$ is applied those systems that were in the lower state, with certain values of A_1 and A_2 as given by (18), now find themselves with the same $A_{1,2}$ which are no longer appropriate to a stationary state with altered Δ. In their new mixed state their dipole moments have a small oscillatory component at frequency

$2E/\hbar$; and the systems initially in the upper state also acquire an oscillatory moment, at the same frequency but vibrating in antiphase. The difference in numbers of systems in the two states leads to an excess contribution from the lower state. The dipole moment after the application of $\mathscr{E}$ therefore begins to oscillate as shown in fig. 3(*a*), starting at the same value as before

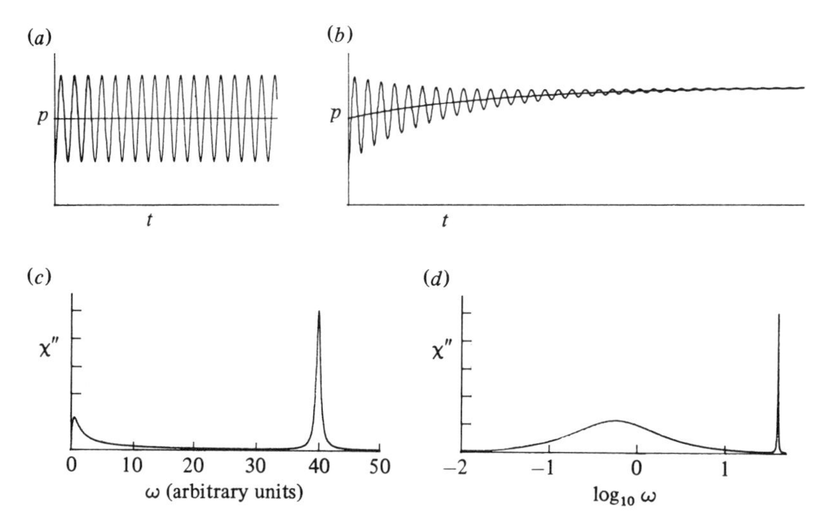

Fig. 3. (*a*) Step-response function of dipole moment of double-well system in the absence of dissipation. (*b*) Step-response function for unequal double-well system with dissipation. (*c*) Lossy part of $\chi(\omega)$ derived from (*b*) according to (64); the amplitudes of the two peaks are not in the correct proportion. (*d*) The same as (*c*) but plotted against $\log \omega$ to show symmetrical form (but markedly different widths) for the relaxation peak on the left and the resonance peak on the right.

$\mathscr{E}$, but now oscillating about a new mean value, which is given by the first term of (23). As time proceeds the oscillation is damped out by radiative processes or randomized by other disturbances, and at the same time the transitions occur to re-equilibrate the assembly and bring in the second term. The latter process, as we shall discuss later, does not normally take place at the same rate as the damping of the oscillations, and the behaviour shown in fig. 3(*b*) may typify the dipole moment response to a step in $\mathscr{E}$ – an exponentially decaying oscillation about a baseline that itself tends exponentially to a constant level.

The physical picture of some systems entirely in one stationary state and the rest in the other is a valuable simplification that was perforce taken for granted in the early days of quantum theory. It gives the correct result when there is no correlation in phase between the occupation amplitudes, a_1 and a_2, since the mean value of cross-terms, e.g. $a_1a_2^*$, is zero and all observables are defined by $a_1a_1^*$ and $a_2a_2^*$. It can lead to error, however, if successive perturbations are close together and insufficient time has elapsed between them for randomization. We shall therefore not use this approximation more than necessary and shall proceed to make the argument of the last paragraph quantitative without invoking its aid until the very end. Let us derive the response of a system, in an arbitrary state, to a step-function $\mathscr{E}$ and to an impulse, the latter being the derivative of the former if the response is linear. For this purpose, and for other problems that follow, the geometrical representation of the state of the system, as in fig. 1, is convenient but needs generalization to include such cases as the unequal double well. For if the poles of the sphere are to continue representing the pure stationary states, account must be taken of their permanent dipole moments.

The interpretation of ζ and ρ in (12) still stands, but the motion of the representative point must be recalculated from (3), with p_2 put equal to $-p_1$, according to (19). Then

$$\dot{a}_1 = (\mathrm{i}\mathscr{E}/\hbar)(p_1 a_1 + p_{12} a_2\, \mathrm{e}^{-\mathrm{i}\omega_0 t})$$

and $$\dot{a}_2 = (\mathrm{i}\mathscr{E}/\hbar)(-p_1 a_2 + p_{12} a_1\, \mathrm{e}^{\mathrm{i}\omega_0 t}),$$

in which $\omega_0 = 2E/\hbar$. It follows that

$$\dot{\zeta} = \dot{a}_2 a_2^* - \dot{a}_1 a_1^* + \text{c.c.} = -2p_{12}\mathscr{E}\eta/\hbar. \tag{24}$$

In evaluating $\dot{\rho}$ it must be remembered that in (12) the phase factors $\mathrm{e}^{\pm Et/\hbar}$ are included in $a_{1,2}$, so that even in the absence of $\mathscr{E}$ the representative point spins at frequency ω_0 round the ζ-axis, i.e. $\dot{\rho} = \mathrm{i}\omega_0\rho$ if $\mathscr{E} = 0$. In the presence of $\mathscr{E}$, (12) shows that

$$\dot{\rho} = \mathrm{i}\omega_0\rho + 2(\dot{a}_1 a_2^* + a_1 \dot{a}_2^*)\,\mathrm{e}^{\mathrm{i}\omega_0 t} = \mathrm{i}(\omega_0 + 2p_1\mathscr{E}/\hbar)\rho + 2\mathrm{i}p_{12}\mathscr{E}\xi/\hbar. \tag{25}$$

The first term reflects the change in rotational speed as E is changed by the interaction of $\mathscr{E}$ with $\pm p_1$. If this rotation is now taken for granted, the remaining changes in the position of the representative point may be collected together from (24) and (25) to give the new equations of motion:

$$\dot{\zeta} = -2p_{12}\mathscr{E}\eta/\hbar, \qquad \dot{\eta} = 2p_{12}\mathscr{E}\xi/\hbar, \qquad \dot{\xi} = 0. \tag{26}$$

These simply describe rotation about the ξ-axis at an angular velocity $2p_{12}\mathscr{E}/\hbar$. The permanent moments $\pm p_1$ in the stationary states play no part in this motion except for the replacement of ω_0 by $\omega_0 + 2p_1\mathscr{E}/\hbar$ in the overall spinning of the pattern about the ρ-axis, and the reduction of p_{12} to a value less than p_0.

Where the interpretation is significantly altered is in the value of $\langle p\rangle$ to be derived from the diagram. According to (10), with the phase factors incorporated in $a_{1,2}$,

$$\langle p\rangle = p_1(a_1 a_1^* - a_2 a_2^*) + p_{12}(a_1 a_2^* + a_1^* a_2) = p_{12}\xi - p_1\zeta. \tag{27}$$

Now (20) allows p_{12} to be written as $p_0 \cos\varepsilon$ and p_1 as $p_0 \sin\varepsilon$, and (19) shows that $\tan\varepsilon = \Delta/\Delta_0$. Hence

$$\langle p\rangle = p_0(\xi \cos\varepsilon - \zeta \sin\varepsilon),$$

and is the ξ-component of the projection of the representative point onto a plane tilted about the y-axis at an angle ε to the horizontal, as in fig. 4(a). The point moves at angular velocity $2E/\hbar$ in a horizontal circular orbit $\mathscr{O}$, when the system is undisturbed; its projection on the tilted plane describes, through its ξ'-component which is the only significant component, an oscillatory dipole moment superposed on a steady component. The

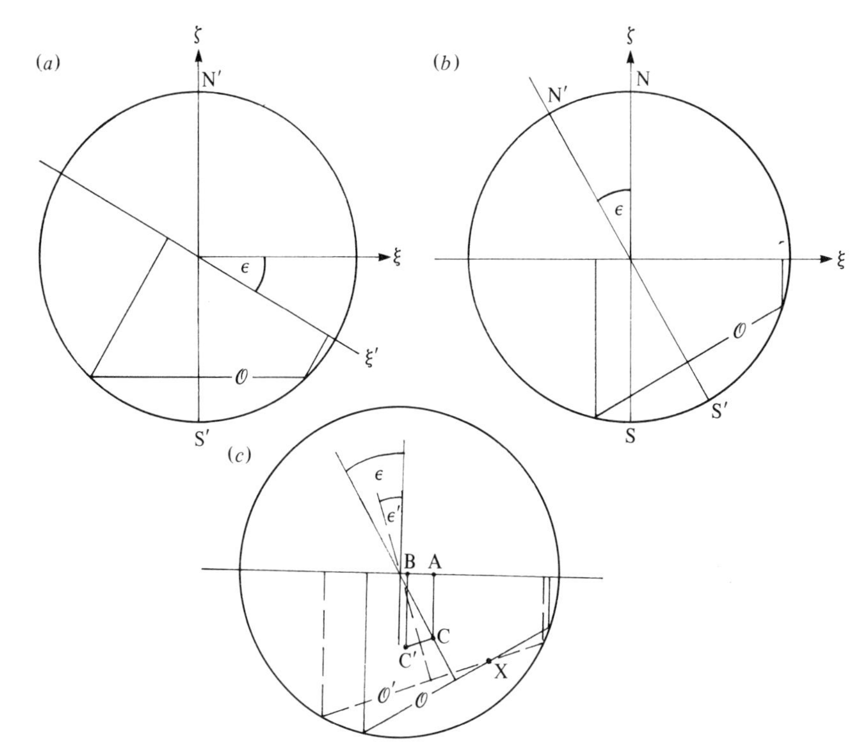

Fig. 4. (a) Modification of fig. 1 appropriate for unequal wells; only the $\xi-\zeta$ section is shown, with $\tan\varepsilon=\Delta/\Delta_0$, and the observed dipole moment is represented by the ξ'-component of the representative point. (b) An alternative form of (a), tilted through ε so that the ξ-component represents the dipole moment, but the orbit of the representative point for the undisturbed system is now tilted. (c) Illustrating the calculation of the step-response function.

latter arises from the permanent moments in the stationary states, still represented by the poles, N' and S'; and is positive if the amplitude of the lower state exceeds that of the upper.

An alternative representation, which has certain advantages, is shown in fig. 4(b), which is just 4(a) tilted through ε so that $\langle p\rangle$ is once more determined by projection onto the horizontal plane. The plane of the orbit $\mathcal{O}$ is now tilted at ε, and N' and S' are the stationary states. This representation amounts to describing each state of the system in terms of the stationary states $\chi_1^{(0)}$ and $\chi_2^{(0)}$ of the equal-well system, in which $\Delta=0$ and $A_1^2=A_2^2=1/2$. Thus the lower stationary state S', in which A_1 and A_2 are given by (18), is a superposition $a_1\chi_1^{(0)}+a_2\chi_2^{(0)}$ with

$$a_1=\tfrac{1}{2}[(1+\Delta/E)^{\frac{1}{2}}+(1-\Delta/E)^{\frac{1}{2}}]$$
$$a_2=\tfrac{1}{2}[(1+\Delta/E)^{\frac{1}{2}}-(1-\Delta/E)^{\frac{1}{2}}],$$

so that, according to (12), $\zeta=-\Delta_0/E=-\cos\varepsilon$ and $\xi=\Delta/E=\sin\varepsilon$; these, of course, are the co-ordinates of S'. It may be noted in passing that in a non-stationary state the amplitudes A_l and A_r in the left-hand and right-hand wells are simply determined by the position of the representative point, since

$$\xi=A_\mathrm{r}A_\mathrm{r}^*-A_\mathrm{l}A_\mathrm{l}^* \quad\text{and}\quad -\zeta+\mathrm{i}\eta=2A_\mathrm{r}^*A_\mathrm{l}. \tag{28}$$

The ξ-coordinate is the difference in occupation of the two wells, which confirms that it represents the dipole moment, while the azimuth angle around the ξ-axis, measured from a point directly below the axis, gives the phase difference between A_r and A_l.

In the representation of fig. 4(b) the response to a step-function $\mathscr{E}$ is immediately obvious. If Δ is the inequality of the wells before $\mathscr{E}$ is applied the point is orbiting at an angle $\tan^{-1}(\Delta/\Delta_0)$; A_l and A_r are not immediately changed by $\mathscr{E}$, but the tilt of the orbit is now altered to $\tan^{-1}(\Delta'/\Delta_0)$, where $\Delta' = \Delta + p_0\mathscr{E}$. In fig. 4($c$) the original orbit $\mathscr{O}$ is shown switching to $\mathscr{O}'$ at X, the position of the point at the instant $\mathscr{E}$ is applied. The change in the mean value and in the oscillatory amplitude of $\langle p \rangle$ can be seen in the projections of the extremes of the orbits onto the horizontal plane. The difference is the step response function, and clearly depends on the position of X. The impulse response is similarly calculated; a strong field $\mathscr{E}$ applied for a short time δt momentarily raises Δ' to a very high value $p_0\mathscr{E}$, and ε rises to $\frac{1}{2}\pi$. The point then rotates about the ξ-axis through $2p_0\mathscr{E}\delta t/\hbar$ before the impulse ceases and the original motion is resumed, but on a slightly displaced orbit. An impulse P_e, defined as $\int \mathscr{E}\,dt$, causes a rotation $2p_0P_e/\hbar$, and the resulting difference in the dipole oscillation, which is P_e times the impulse response function, again clearly depends on the moment of application of the impulse.

In an assembly of randomly phased systems, therefore, different systems will respond differently. But the response of the mean is easily discovered. Randomness of phasing implies that before a step is applied the centroid of all points representing the different members of the assembly lies on the ζ'-axis, N′S′ in fig. 4(b), and that the mean of all the oscillatory dipole moments vanishes, though there will in general be a steady moment since except at a very high temperature the centroid lies below the centre of the sphere. When application of $\mathscr{E}$ causes each point to orbit about a new axis, the centroid does the same. Thus if C in fig. 4(c) is the original centroid, the mean response will be the orbit CC′, whose projection gives the response $\bar{p}(t)$; it is clear that, as in fig. 3, $\bar{p}$ starts at one extreme of its cycle of oscillation. If $\mathscr{E}$ is small, the angle $\varepsilon' - \varepsilon$ is $(\mathrm{d}/\mathrm{d}\Delta)\tan^{-1}(\Delta/\Delta_0)\cdot p_0\mathscr{E}$, i.e. $\Delta_0 p_0\mathscr{E}/E^2$; a short calculation shows that if f is the length OC, the fractional excess in the lower state, the oscillatory amplitude, $\frac{1}{2}\mathrm{AB}\times p_0$, is $(\Delta_0 p_0^2\mathscr{E}f/E^2)\cos\varepsilon$, or $\Delta_0^2 p_0^2\mathscr{E}f/E^3$, in agreement with the first term of (23) when f takes the thermal equilibrium value $\tanh(E/k_BT)$. The step-response function, when $\mathscr{E} = 1$, is written down by inspection:

$$S(t) = (\Delta_0^2 p_0^2 f/E^3)[1 - \cos(2Et/\hbar)], \tag{29}$$

and its derivative $\mathrm{d}S/\mathrm{d}t$ is the impulse-response function:

$$h(t) = (2\Delta_0^2 p_0^2 f/\hbar E^2)\sin(2Et/\hbar). \tag{30}$$

This is just what results from turning the point C through $2p_0/\hbar$ about the ξ-axis, as prescribed above. So long as the displacements of C are small, linearity holds and the effects of successive impulses are additive. These expressions take no account of dissipative processes and therefore offer no explanation of the second term in (23). This will arise out of the next development of the theory.

Radiative decay of the two-level system

We shall begin by solving a model from which the required results can be derived without much additional calculation, by building on the ideas developed in chapter 16. The model consists of a two-level system for a particle of mass m_1 tunnelling between equal wells, and a harmonic oscillator of mass m_0 and frequency ω_0, close to $2\Delta_0/\hbar$. The two are weakly coupled as if an electric field proportional to x, the displacement of the oscillator, acted to polarize the double well; conversely the double well acts on the oscillator with a force of constant magnitude but whose sign depends on which well the particle occupies. The coupling term is incorporated in the Hamiltonian for the combined system: **66**

$$\mathcal{H}(x, y) = T_0(\dot{x}) + V_0(x) + T_1(\dot{y}) + V_1(y) + Cx \operatorname{sgn}(y), \tag{31}$$

in which T_1 and V_1 refer to the double well, T_0 and V_0 to the oscillator, and $\operatorname{sgn}(y) = \pm 1$ according as y is positive or negative. Correspondingly Schrödinger's time-independent equation takes the form

$$(\hbar^2/2m_1)\partial^2\Phi/\partial y^2 + (\hbar^2/2m_0)\partial^2\Phi/\partial x^2 + [W - V_1 - V_0 - Cx \operatorname{sgn}(y)]\Phi = 0. \tag{32}$$

If $C = 0$ the solutions, of energy W_0, are product wave-functions, $\chi(y)\psi(x)$, of which the components are of either even parity ($\chi_1(y)$, $\psi_{2l}(x)$) or odd ($\chi_2(y)$, $\psi_{2l+1}(x)$). The coupling term is odd in both x and y, and the matrix element connecting two product wave-functions vanishes unless one contains χ_1 and the other χ_2; moreover, since the $\psi(x)$ are oscillator functions, the two appearing in any non-vanishing matrix element must be neighbours, ψ_l and ψ_{l+1}. With weak coupling, then, every $\chi_1\psi_n$ suffers an admixture of $\chi_2\psi_{n\pm1}$, and every $\chi_2\psi_n$ an admixture of $\chi_1\psi_{n\pm1}$. The mixing is strongest when the energy levels involved are close together; thus when $\hbar\omega_0 \approx 2\Delta_0$, $\chi_1\psi_{n+1}$ is most strongly modified by the admixture of $\chi_2\psi_n$, and $\chi_2\psi_n$ has the strongest admixture of $\chi_1\psi_{n+1}$. We therefore seek real solutions of (32) with the form:

$$\Phi_n = \alpha_n\chi_1\psi_{n+1} + \beta_n\chi_2\psi_n; \qquad \alpha_n^2 + \beta_n^2 = 1. \tag{33}$$

Substituting in (32) we have that

$$\alpha_n[W + \Delta_0 - E_{n+1} - Cx \operatorname{sgn}(y)]\chi_1\psi_{n+1} + \beta_n[W - \Delta_0 - E_n - Cx \operatorname{sgn}(y)]\chi_2\psi_n = 0, \tag{34}$$

in which $E_n = (n + \frac{1}{2})\hbar\omega_0$. Small terms omitted through the approximations in (33) give a negligible contribution when (34) is multiplied by either $\chi_1\psi_{n+1}$ or $\chi_2\psi_n$ and integrated over both variables to yield the first-order approximations:

$$\alpha_n(\Omega_n+\omega_e)=\beta_n CM^{(1)}_{n,n+1}/\hbar \quad \text{and} \quad \beta_n(\Omega_n-\omega_e)=\alpha_n CM^{(1)}_{n,n+1}/\hbar, \tag{35}$$

in which $\Omega_n = W/\hbar-(n+1)\omega_0$, $\omega_e=\Delta_0/\hbar-\frac{1}{2}\omega_0$, and $M^{(1)}_{n,n+1}$ is the same matrix element as in (13.26) except that in laboratory units its value is $[\hbar(n+1)/2m_0\omega_e]^{\frac{1}{2}}$.

I.366 Once again we have a close analogy to the coupled circuits of chapter 12, the energy level structure obtained by eliminating α_n and β_n from (35) having the same hyperbolic form as the normal mode frequency structure of (12.11):

$$\Omega_n^2=\omega_e^2+\gamma^2(n+1), \quad \text{where } \gamma=C/(2\hbar\omega_0 m_0)^{\frac{1}{2}}. \tag{36}$$

When the systems are uncoupled the mismatch between ω_0 and $2\Delta_0/\hbar$ causes the levels to appear in pairs, as shown on the left of fig. 5. Coupling

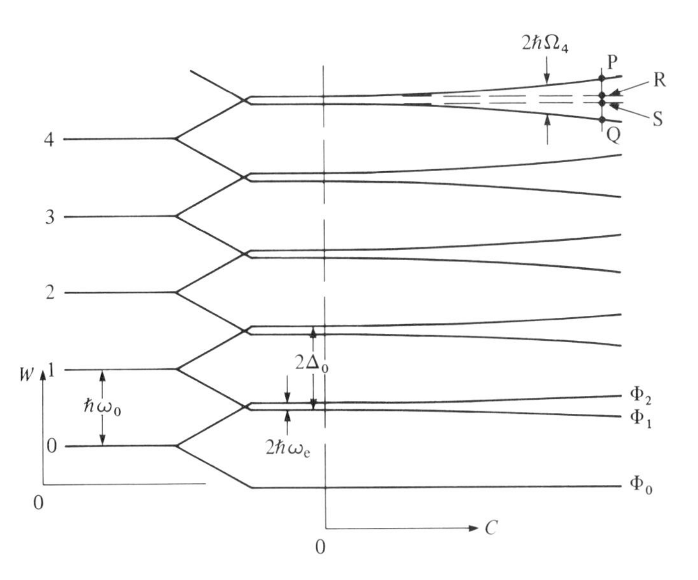

Fig. 5. Energy levels for a double-well system coupled to a harmonic vibrator. On the left, where the systems are uncoupled, the levels have energies $(n+\frac{1}{2})\hbar\omega_0\pm\Delta_0$, and in the diagram $2\Delta_0$ is shown as slightly greater than $\hbar\omega_0$. With increase of coupling constant C the splitting increases, more for the more excited vibrator levels.

enlarges the separation, and the presence of $(n+1)$ in (36) enhances the effect at the higher levels; the solitary lowest level is unperturbed. It should be remembered that this analysis goes only as far as first-order effects of C on the wave-function (second-order in energy), so that by the time the splitting is as large as on the right of the diagram higher-order effects will certainly begin to play a part and change the details. We shall treat only first-order effects. From (35),

$$\alpha_n^2/\beta_n^2=(\Omega_n-\omega_e)/(\Omega_n+\omega_e). \tag{37}$$

In the diagram ω_e is taken to be positive, and the state P has $(\Omega_n-\omega_e)$ represented by PR, $(\Omega+\omega_e)$ by PS. Hence $\alpha_n/\beta_n=(\text{PR}/\text{PS})^{\frac{1}{2}}$ and (35) shows that α_n and β_n have the same sign. The state Q has the ratio $|\alpha_n/\beta_n|$ inverted, and opposite signs for α_n and β_n. This behaviour exactly parallels what has already been noted in analogous cases. Since $\alpha_n^2+\beta_n^2=1$, it follows from (37) that in the state P

$$\alpha_n^2=\tfrac{1}{2}(1-\omega_e/\Omega_n), \qquad \beta_n^2=\tfrac{1}{2}(1+\omega_e/\Omega_n). \tag{38}$$

Let us use this result to follow the behaviour of the two-well system very weakly coupled to the unexcited oscillator. This will serve as a model for the radiation of energy into one mode of a cavity. As in chapter 16 we shall suppose the cavity to be so large that at any instant after coupling the chance of finding any one mode excited is extremely small. This places a helpful restriction on the form of the wave-function which will contain only the lowest three states of the combined system, Φ_0, Φ_1, and Φ_2 as shown in fig. 5. At the instant of coupling, taken as $t=0$ for convenience, the spatial part Ψ may be written

$$\left.\begin{aligned}&\Psi(0)=\lambda_0\Phi_0+\lambda_1\Phi_1+\lambda_2\Phi_2,\\ \text{where}\quad &\Phi_0=\chi_1\psi_0, \qquad \Phi_1=-\beta_0\chi_1\psi_1+\alpha_0\chi_2\psi_0,\\ &\Phi_2=\alpha_0\chi_1\psi_1+\beta_0\chi_2\psi_0 \quad\text{and}\quad \lambda_0\lambda_0^*+\lambda_1\lambda_1^*+\lambda_2\lambda_2^*=1.\end{aligned}\right\} \tag{39}$$

Hence $\Psi(0)=\lambda_0\chi_1\psi_0+(\lambda_1\alpha_0+\lambda_2\beta_0)\chi_2\psi_0+(\lambda_2\alpha_0-\lambda_1\beta_0)\chi_1\psi_1,$ (40)

and since we take the oscillator as unexcited the coefficient of ψ_1 must vanish:

i.e. $$\lambda_2\alpha_0=\lambda_1\beta_0. \tag{41}$$

Absolute phase being irrelevant, we shall take λ_1 and λ_2 to be real; λ_0 may be complex.

The presence of three terms in $\Phi(0)$ shows that subsequently Ψ will exhibit a rapidly oscillatory behaviour, at frequency ω_0, modulated by the slow beat due to Φ_1 and Φ_2, at frequency Ω_0. Correspondingly the energy of the double-well system and the amplitude of the oscillatory dipole moment will fluctuate at the low frequency Ω_0. At time t,

$$\Psi(t)=\lambda_0\,e^{i\omega_e t}\,\Phi_0+\lambda_1\,e^{-i(\omega_0-\Omega_0)t}\,\Phi_1+\lambda_2\,e^{-i(\omega_0+\Omega_0)t}\,\Phi_2$$

which may be written, like (40):

$$\begin{aligned}\Psi(t)=\lambda_0\,e^{i\omega_e t}\chi_1\psi_0&+(\lambda_1\alpha_0\,e^{i\Omega_0 t}+\lambda_2\beta_0\,e^{-i\Omega_0 t})\,e^{-i\omega_0 t}\chi_2\psi_0\\ &+(\lambda_2\alpha_0\,e^{-i\Omega_0 t}-\lambda_1\beta_0\,e^{i\Omega_0 t})\,e^{-i\omega_0 t}\chi_1\psi_1.\end{aligned} \tag{42}$$

The probability $\mathscr{P}_2$, of finding the double-well system in its excited state is determined by the second term, the only one to contain χ_2:

$$\begin{aligned}\mathscr{P}_2(t)=|\lambda_1\alpha_0\,e^{i\Omega_0 t}+\lambda_2\beta_0\,e^{-i\Omega_0 t}|^2-(\lambda_1\alpha_0+\lambda_2\beta_0)^2-4\lambda_1\lambda_2\alpha_0\beta_0\sin^2\Omega_0 t\\ =\mathscr{P}_2(0)[1-(\gamma^2/\Omega_0^2)\sin^2\Omega_0 t],\end{aligned} \tag{43}$$

by use of (36), (38) and (41).

The exchange of energy implied by (43) has exactly the same form as the radiative exchange between an oscillator and a cavity mode, as described by (16.14), if the coupling is so weak that Ω_0 may be put equal to ω_e. As in the earlier example, we have made here no assumptions about the initial state, and its parameters have vanished from (43); it follows that coupling to all modes in an empty cavity causes $\mathscr{P}_2$ and the excitation energy of the system to decay exponentially. Since (43) shows that the initial energy transfer to a single mode follows the same behaviour as (16.16), i.e. $E(0)-E(t)=E(0)\gamma^2t^2$, we may proceed immediately to determine τ_e, the time constant for energy loss into a cavity. Equating γ^2 to $1/2\pi g\tau_e$, we have that

$$\tau_e = 1/2\pi g\gamma^2, \tag{44}$$

where $g = 4\Delta_0^2 V/\pi^2\hbar^2c^3$, the density of states for a cavity of volume V at a frequency of $2\Delta_0/\hbar$.

To find the value of γ we go back to the Hamiltonian (31). Consider a typical mode at frequency $2\Delta_0/\hbar$, and at some instant when the field $\mathscr{E}$ is at its peak write $\overline{\mathscr{E}^2}$ for the mean square field, averaged over the cavity. Then the mode energy, $\frac{1}{2}\varepsilon_0 V\overline{\mathscr{E}^2}$, must be equated to $\frac{1}{2}m_0(2\Delta_0/\hbar)^2x^2$, the potential energy of the equivalent oscillator; and the square of the coupling energy, C^2x^2 in (31), must be equated to $\frac{1}{3}p_0^2\overline{\mathscr{E}^2}$, the factor 1/3 arising as in chapter 16 from the different polarizations of different modes. Hence 73

$$C^2 = 4m_0\Delta_0^2p_0^2/3\hbar^2\varepsilon_0 V, \tag{45}$$

and

$$\gamma^2 = \Delta_0 p_0^2/3\hbar^2\varepsilon_0 V, \tag{46}$$

from (36), with $\hbar\omega_0$ put equal to $2\Delta_0$.

From (44), then,

$$\tau_e = 3\pi\hbar^4 c/8\mu_0\Delta_0^3p_0^2. \tag{47}$$

Now a Hertzian dipole of mass m_{12}, oscillating at $2\Delta_0/\hbar$, has a time-constant for energy loss of $3\pi m_{12}\hbar^2c/2\mu_0\Delta_0^2e^2$. These time-constants coincide if f_{12}, i.e. m/m_{12}, is $4m\Delta_0p_0^2/\hbar^2e^2$, which is the value already calculated directly. 73 99 The double-well system not only responds to weak stimulation as if it were a harmonic oscillator, but it also radiates spontaneously like the same equivalent oscillator. Of course, if we had not reached this conclusion it would have been very disturbing, for the Einstein argument involving detailed balance between the radiation field and a material oscillator (to which we revert shortly) demands that a unique relation, as expressed by (16.29), between stimulated and spontaneous processes must hold if the Boltzmann distribution is to apply universally.

When the motion of a classical oscillator decays exponentially the time-constant for energy decay is half that for the decay of amplitude, i.e. dipole moment, and the same holds for the double-well system. To show this, we return to (42) and evaluate $\langle y(t)\rangle$, i.e. $\iint \Psi^*(t)y\Psi(t)\,dy\,dx$. Of the nine terms in the product $\Psi^*\Psi$, by symmetry only two survive, those which have the form $\chi_1\chi_2\psi_0^*\psi_0$; with the help of (41) we find:

$$\langle p\rangle/p_0 = 2\,\mathrm{Re}\,[(\lambda_0^*\lambda_1/\alpha_0)\,e^{-i(\omega_0+\omega_e)t}(\alpha_0^2\,e^{i\Omega_0 t}+\beta_0^2\,e^{-i\Omega_0 t})]. \tag{48}$$

The first part, including $e^{-i(\omega_0+\omega_e)t}$, is a fast oscillation whose amplitude is slowly modulated by the following term in brackets, and it is the absolute magnitude of this latter term, and its variation with time, that carries the information needed to determine the decay of the dipole oscillation. If we write Amp (p) for $|\alpha_0^2 e^{i\Omega_0 t}+\beta_0^2 e^{-i\Omega_0 t}|$, we have that

$$[\text{Amp}\,(p)]^2 = 1-4\alpha_0^2\beta_0^2 \sin^2 \Omega_0 t = 1-(\gamma^2/\Omega_0^2)\sin^2 \Omega_0 t,$$

which has the same form as (43). We conclude that the square of the oscillatory dipole moment decays exponentially with time-constant τ_e, and consequently the moment itself decays with time-constant τ_p equal to $2\tau_e$, like a classical dipole.

Decay by spontaneous radiation is readily incorporated in the geometrical representation of the double-well system. In the absence of external perturbations the representative point spins at angular frequency $2E/\hbar$ about the tilted axis of fig. 4(b)†, but now its height above the lower state S′, measured along S′N′, decreases exponentially with time-constant τ_e, since this is a measure of its excitation energy, while the radius of its circular orbit, which determines the dipole oscillations, decreases at half the rate. The necessity of allowing the point to leave the surface of the sphere illustrates a basic feature of quantum mechanics, that while the wave-function of two or more uncoupled systems may be written as a simple product, once they are coupled the function evolves so that in general it can never again be written thus, even after the coupling is reduced to zero; only by making observations on the state of each component separately can the product wave-function be restored. In this case the representative point, once located on the sphere, must remain there so long as the system is truly isolated from other quantal systems, since the energy and dipole amplitude are then uniquely related. Coupling to the cavity destroys the uniqueness of the relation.

Equations of motion of a damped double-well

When the double-well is subject to weak external forces as well as to radiative damping the two processes may be treated as independent in any short interval. The proviso that the force shall be weak is necessary because τ_e and τ_p vary with Δ. To find the variation, note that according to (16.11) a Hertzian dipole of mass m_{12} has a decay time proportional to m_{12}/ω_{12}^2. Now (17.2) and (2) show that f_{12} here is $2m\omega_{12}p_{12}^2/\hbar e^2$, varying as $1/E$ according to (19), since $\omega_{12}\propto E$. Hence τ_e and τ_p, being proportional to $1/\omega_{12}^2 f_{12}$, vary as $1/E$. Relative to the equal well system, the decay times must be reduced by the factor Δ_0/E, i.e. $\Delta_0/(\Delta_0^2+p_0^2\mathscr{E}^2)^{\frac{1}{2}}$. Moreover when $p_0\mathscr{E}$ is comparable to Δ_0 the representation of fig. 4(b) and (c) shows that decay takes place along a tilted axis. If the force is imagined to be applied

† Here we generalize to an unequal double-well the results derived only for the equal wells.

as a sequence of sharp impulses, decay is fast during the application of the impulses. It is readily shown, however, that provided the mean value of $p_0\mathscr{E}$ is much less than Δ_0 these effects may be neglected, and the decay assumed to take place only between the impulses, and at the normal speed. We shall now make this assumption in deriving the impulse-response function of a radiatively damped double-well.

At any instant the wave-function may be expressed as a sum over stationary state solutions for the uncoupled systems, which are product wave-functions of the type $\chi_1\psi_n(x_i)$ or $\chi_2\psi_n(x_i)$, $\psi_n(x_i)$ being a shorthand notation for product wave-functions comprising one oscillator function for each cavity mode. Immediately before the impulse

$$\Psi = \sum_n [\lambda_n\chi_1\psi_n(x_i) + \mu_n\chi_2\psi_n(x_i)]. \tag{49}$$

A weak impulse, $\int \mathscr{E}\,\mathrm{d}t = P$, acting on the double-well only, adds on to Ψ an increment $ieyP\Psi/\hbar$, according to (13.31). Now χ_1 has amplitudes $+1/\sqrt{2}$ in each well when they are equal (the only case we shall consider) and χ_2 has amplitudes $\pm 1/\sqrt{2}$. The impulse adds $\mathrm{i}p_0P/\sqrt{2}\hbar$ to the amplitude in the right-hand well of χ_1, and subtracts the same from the left-hand well – that is to say, χ_1 is converted into $\chi_1 + (\mathrm{i}p_0P/\hbar)\chi_2$; similarly χ_2 is converted into $\chi_2 + (ip_0P/\hbar)\chi_1$. Every λ_n in (49) is thus converted into $\lambda_n + \mathrm{i}\kappa\mu_n$ and every μ_n into $\mu_n + \mathrm{i}\kappa\lambda_n$, κ being written for $p_0P/\hbar$.

The initial probability $\mathscr{P}_1$ of finding the double-well system in its ground state, whatever the oscillators may be doing, is $\Sigma_n\lambda_n\lambda_n^*$, and the probability $\mathscr{P}_2$ that it will be found excited is $\Sigma_n\mu_n\mu_n^*$. The change in each λ_n and μ_n due to the impulse results in a first-order change to $\mathscr{P}_1$ and $\mathscr{P}_2$:

$$\delta\mathscr{P}_1 = -\delta\mathscr{P}_2 = \mathrm{i}\kappa\sum_n(\mu_n\lambda_n^* + \mu_n^*\lambda_n). \tag{50}$$

By forming $\langle\Psi^*|y|\Psi\rangle$ the dipole moment is calculated:

$$\langle p\rangle/p_0 = \sum_n(\mu_n\lambda_n^* + \mu_n^*\lambda_n). \tag{51}$$

These results immediately suggest the appropriate modification to the geometrical representation expressed by (12):

$$\zeta = \sum_n(\mu_n\mu_n^* - \lambda_n\lambda_n^*) = \sum_n(1 - 2\lambda_n\lambda_n^*), \qquad \rho = \xi + \mathrm{i}\eta = 2\sum_n\lambda_n\mu_n^*. \tag{52}$$

It is not difficult to verify that no point specified in this manner can lie outside the unit sphere, and that to lie on the sphere $\mu_n = c\lambda_n$, c being a complex constant so that (49) takes the form appropriate to uncoupled system and cavity, $\Psi = (\chi_1 + c\chi_2)\Sigma_n\lambda_n\Psi_n$. Otherwise the point lies within the sphere, as we have come to expect from the analysis of coupled systems. The prescription (52) matches the requirement that $\frac{1}{2}(\zeta+1)$ shall describe the probability of finding the double-well system excited, and that ξ shall describe the dipole moment. Further, when unperturbed the representative point executes an orbit round the ζ-axis with angular velocity $2\Delta_0/\hbar$, since each term in ρ has its phase angle increasing at this rate. We may be satisfied, therefore, that the dissipative process will take place as already determined, with time-constant τ_e for $1+\zeta$, $2\tau_e$ for ρ, and we can also write down the response to an impulse:

$$\delta\rho = 2\sum_n (\lambda_n \delta\mu_n^* + \mu_n^* \delta\lambda_n) = 2i\kappa \sum_n (\mu_n \mu_n^* - \lambda_n \lambda_n^*) = 2i\kappa\zeta.$$

$$\therefore \quad \delta\xi = 0, \qquad \delta\eta = 2\kappa\zeta, \quad \text{and similarly } \delta\zeta = -2\kappa\eta.$$

This describes rotation through 2κ, i.e. $2p_0P/\hbar$, about the ξ-axis, exactly the same as found in (30) for the isolated double-well.

We may now write the equations of motion, under the influence of a varying field $\mathscr{E}(t)$, in Cartesian co-ordinates:

$$\left.\begin{aligned} \dot{\xi} &= 2\Delta_0\eta/\hbar - \xi/\tau_p \\ \dot{\eta} &= -2\Delta_0\xi/\hbar - \eta/\tau_p + 2p_0\mathscr{E}\zeta/\hbar \\ \dot{\zeta} &= (\zeta_0 - \zeta)/\tau_e - 2p_0\mathscr{E}\eta/\hbar. \end{aligned}\right\} \qquad (53)$$

The three contributions, rotation about ζ at frequency $2\Delta_0/\hbar$, rotation about ξ due to $\mathscr{E}$, and dissipation with two separate time-constants, are immediately recognizable in the structure of these equations. The opportunity has been taken to replace $-(1+\zeta)/\tau_e$ in the third equation by $(\zeta_0-\zeta)/\tau_e$, so that the ζ-coordinate is permitted to relax exponentially towards a point which is not the pure lower state. This, as will be justified shortly, allows for stimulated as well as spontaneous processes at a non-zero temperature.

These equations of motion were developed for the equal-well case and are virtually the same as the Bloch equations for nuclear spin resonance, (8.18) and (8.19). The two time-constants in (53) are counterparts of T_1 (longitudinal relaxation time $\equiv \tau_e$) and T_2 (transverse relaxation time $\equiv \tau_p$). In general there are other dissipative mechanisms besides radiative transitions, and τ_p need not be $2\tau_e$ – in fact, as we discuss later, energy-conserving dephasing processes can lower τ_p without changing τ_e; it is not possible for τ_p to exceed $2\tau_e$. When the wells are unequal or $\mathscr{E}$ is strong, or both together, modifications must be made to the axes and speeds of rotation, and to the decay times; in the representation of fig. 4(*b*) the tilt, ε, is determined by the instantaneous value of E, to which both Δ and $p_0\mathscr{E}$ contribute, and the tilted axis is also the direction of energy relaxation with τ_e replaced by $\Delta_0\tau_e/E$. Clearly if $\mathscr{E}$ is both strong and variable the equations become unpleasantly non-linear, and we shall not even write them down.

I.227

When τ_e and τ_p are long and $\mathscr{E}$ oscillates at a frequency close to $2\Delta_0/\hbar$, even a weak field can cause considerable perturbation. The behaviour illustrated in fig. 1 is characteristic of a non-dissipative system, and is essentially transient when dissipation is present. Let us look for the steady-state solution of (53), once this transient has died away and the variations of ξ, η and ζ are synchronized to the driving frequency, ω. If $\mathscr{E} = \mathscr{E}_0 \cos \omega t = \frac{1}{2}\mathscr{E}_0(e^{i\omega t} + e^{-i\omega t})$, the first two equations show that ξ and η are proportional to $\mathscr{E}$ in the lowest order of approximation, so that the third equation contains $\mathscr{E}$ only as a quadratic term. Let us assume, then, that ζ settles down to a constant value, $\bar{\zeta}$, which is not ζ_0 since there is a steady term contained in $\mathscr{E}\eta$. With ζ constant the first two equations are conveniently written in terms of $\rho(=\xi+\mathrm{i}\eta)$:

$$\tau_p\dot{\rho}+(1+2\mathrm{i}\Delta_0\tau_p/\hbar)\rho=(\mathrm{i}p_0\tau_p\bar{\zeta}\mathscr{E}/\hbar)(\mathrm{e}^{\mathrm{i}\omega t}+\mathrm{e}^{-\mathrm{i}\omega t}). \tag{54}$$

Near resonance the second exponential dominates the solution if $2\Delta_0\tau_p\gg 1$, and we may neglect the positive exponential. Then $\rho=\rho_0\,\mathrm{e}^{-\mathrm{i}\omega t}$, where

$$\rho_0=(\mathrm{i}p_0\mathscr{E}_0\tau_p\bar{\zeta}/\hbar)/[1+\mathrm{i}(\omega_{12}-\omega)\tau_p]; \qquad \omega_{12}=2\Delta_0/\hbar. \tag{55}$$

The imaginary part of ρ_0 is the amplitude of η, from which it follows that $\eta=(p_0\mathscr{E}_0\tau_p\bar{\zeta}/\hbar)[\cos\omega t-(\omega_{12}-\omega)\tau_p\sin\omega t]/[1+(\omega_{12}-\omega)^2\tau_p^2]$, and the mean value of $\mathscr{E}\eta$ is $\frac{1}{2}p_0\mathscr{E}_0^2\tau_p\bar{\zeta}/\hbar[1+(\omega_{12}-\omega)^2\tau_p^2]$. When this is substituted in the third equation of (53), and $\dot{\zeta}$ put equal to zero, the mean value of ζ emerges as the solution:

$$\bar{\zeta}=\zeta_0\Big/\left[1+\frac{p_0^2\mathscr{E}_0^2\tau_e\tau_p/\hbar^2}{1+(\omega_{12}-\omega)^2\tau_p^2}\right]. \tag{56}$$

If ζ_0 is the equilibrium position of the unperturbed system maintained in contact with black-body radiation at temperature T, $\zeta_0=-\tanh(\Delta_0/k_BT)$ and the reduction in $|\zeta_0|$ expressed by (56) can be interpreted as a raising of the effective temperature, T_{eff}. The energy input by virtue of $\mathscr{E}$ acting on the oscillating dipoles can only be dissipated at a rate $(\bar{\zeta}-\zeta_0)/\tau_e$ times the excess. The reason why τ_p appears in (56) as well as τ_e is that it helps determine the magnitude of the oscillating dipole and through this the energy input. If τ_e and τ_p are long enough, and $\omega_{12}-\omega$ small enough, $\bar{\zeta}$ may be reduced practically to zero ($T_{\mathrm{eff}}\to\infty$) while $\mathscr{E}$ is still small enough for the linear approximation to apply.

The strength of the dipole response follows from (55):

$$|\rho_0|=(p_0\mathscr{E}_0\tau_p\zeta_0/\hbar)[1+(\omega_{12}-\omega)^2\tau_p^2]^{\frac{1}{2}}/[1+(\omega_{12}-\omega)^2\tau_p^2+p_0^2\mathscr{E}_0^2\tau_p\tau_e/\hbar^2]. \tag{57}$$

So long as $\mathscr{E}_0$ is weak enough to cause insignificant 'heating' $|\rho_0|\propto\mathscr{E}_0$, but as $\mathscr{E}_0$ is increased $|\rho_0|$ ultimately falls to zero as $1/\mathscr{E}_0$. Naturally the saturation of $|\rho_0|$ is most readily achieved at resonance, when $\omega=\omega_{12}$.

At resonance $\bar{\zeta}$, according to (56), is reduced by a factor $1+2p_0^2\mathscr{E}_0^2\tau_e^2/\hbar^2$, if $\tau_p=2\tau_e$. Writing $\hbar e^2f/2\omega_{12}m$ for p_0^2, $6\pi mc/f\mu_0\omega_{12}^2e^2$ for τ_e and $2\Phi/\varepsilon_0c$ for $\mathscr{E}_0^2$, Φ being the power flux in the irradiation, we have that for a transition at a wavelength λ,

107

$$2p_0^2\mathscr{E}_0^2\tau_e^2/\hbar^2=9m\lambda^5\Phi/4\pi^3\mu_0c^2e^2\hbar\doteqdot 2\times10^{29}\lambda^5\Phi/f.$$

A typical optical transition, with $f\sim1$ and $\lambda\sim5\times10^{-7}$ m, requires a flux of about 160 $W\,\mathrm{m}^{-2}$ to reduce ζ_0 by a factor 2 and thus go some way towards saturating the transition. This is a flux density that is readily available in a monochromatic beam by use of a continuous gas laser. At lower frequencies, as for example in the microwave range of the spectrum, the dependence of Φ on λ^{-5}, to give a certain degree of saturation, would suggest that even the weakest sources would saturate the transitions. It should be remembered, however, that this calculation takes note only of radiative processes in dissipation, and it is the inefficiency of these at long wavelengths that produces this result. In practice collisions between the molecules dominate the loss mechanisms, but even so it is not hard to saturate the transitions with very modest irradiating power. The same applies to nuclear magnetic resonance where very low power levels are normally needed.

Stimulated and spontaneous transitions

The introduction of ζ_0 in (53), instead of -1 as required by the theory of spontaneous transitions, is a fairly obvious generalization, but one which deserves discussion. Let us imagine the double-well system immersed in a bath of cavity radiation in equilibrium at temperature T. Following our usual practice we consider the interaction with one cavity oscillator, but now it is not initially in its ground state, but in some mixture of excited states, so that at the instant of coupling the product wave-function takes the form:

$$\Phi(0) = (a_1\chi_1 + a_2\chi_2)\sum_n b_n\psi_n. \tag{58}$$

If we are concerned with the average behaviour of the energy contained in an assembly of similar systems we may assume not only that the cavity oscillators have random phases for their ψ_n, but that the only results of interest are those obtained by averaging over all phases of a_1 and a_2. This allows us to follow the development of each term in (58) separately (which the linearity of Schrödinger's equation permits) and to ignore phase correlations when combining the results. Thus, we start with a typical term, $a_1\chi_1 b_n\psi_n$, and express it as a sum of stationary states (33):

$$\chi_1\psi_n = \alpha_{n-1}\Phi' + \beta_{n-1}\Phi'', \tag{59}$$

in which $$\Phi' = \alpha_{n-1}\chi_1\psi_n + \beta_{n-1}\chi_2\psi_{n-1}$$

and $$\Phi'' = \beta_{n-1}\chi_1\psi_n - \alpha_{n-1}\chi_2\psi_{n-1}.$$

The subsequent development of the wave-function results from the beating of the two terms in (59) with frequency $2\Omega_{n-1}$. After half a cycle the relative phases are reversed, and $\chi_1\psi_n$ is turned into $(\alpha_{n-1}^2 - \beta_{n-1}^2)\chi_1\psi_n + 2\alpha_{n-1}\beta_{n-1}\chi_2\psi_{n-1}$. The probability $\mathcal{P}_2$ of finding the system excited is now $4\alpha_{n-1}^2\beta_{n-1}^2$, i.e., $\gamma^2 n/\Omega_{n-1}^2$ from (38) and (36). At any other time

$$\mathcal{P}_2(t) = (\gamma^2 n/\Omega_{n-1}^2)\sin^2(\Omega_{n-1}t) \propto nt^2 \text{ for small } t.$$

Proceeding along the same lines as for the unexcited cavity we conclude that if the system is initially unexcited $\mathscr{P}_2$ will start to rise linearly with time, the rate being determined by the mean value of n for those oscillators which are in resonance with the system.

If, however, we proceed likewise with an initial state $a_2\chi_2 b_n\psi_n$ the argument is similar except that the analogue of (59) contains ψ_n and ψ_{n+1} rather than ψ_{n-1} and ψ_n. The probability $\mathscr{P}_1$, of finding the system unexcited again rises linearly, but the rate is controlled now by the mean value of $n+1$, rather than n.

The development of $\mathscr{P}_1$ and $\mathscr{P}_2$ for the assembly is now clear, for randomness of phase-relationships eliminates any interference effects, so that the two processes make independent contribution to the evolution of $\mathscr{P}_1$ and $\mathscr{P}_2$;

$$-\dot{\mathscr{P}}_2 = \dot{\mathscr{P}}_1 = -\bar{n}\mathscr{P}_1/\tau_e + \overline{(n+1)}\mathscr{P}_2/\tau_e = (\bar{\mathscr{P}}_1 - \mathscr{P}_1)/\tau_e', \tag{60}$$

in which $\tau_e' = \tau_e/(2n+1)$ and τ_e is the decay time for spontaneous radiation into the empty cavity. The equilibrium value of $\mathscr{P}_1$ is $\bar{\mathscr{P}}_1$, where

$$\bar{\mathscr{P}}_1 = \frac{\bar{n}+1}{2\bar{n}+1} \quad \text{and} \quad \bar{\mathscr{P}}_2 = 1 - \bar{\mathscr{P}}_1 = \frac{\bar{n}}{2\bar{n}+1}. \tag{61}$$

Since according to Planck's law, $\bar{n} = (e^{2\Delta_0/k_BT} - 1)^{-1}$,

$$\bar{\mathscr{P}}_1/\bar{\mathscr{P}}_2 = 1 + 1/\bar{n} = e^{2\Delta_0/k_BT},$$

as required by the Boltzmann distribution. This is the justification for placing ζ_0 instead of -1 in (53), but it should also be noted that the stimulated processes resulting from the excited cavity modes speed up the process of equilibration, and that τ_e in (53) should be multiplied by $1/(2\bar{n}+1)$, i.e. $\tanh(\Delta_0/k_BT)$. The same holds for τ_p which remains equal to $2\tau_e$, although to prove this it would be necessary to take account of phase correlation between a_1 and a_2, since otherwise there would be no dipole oscillation to decay.

The dependence of τ_e on $\bar{n}$, and thus on the ambient temperature, contrasts with the constancy of τ_e for a harmonic vibrator. In fact the transition rate between any two levels is affected by temperature in exactly the same way in both systems, but as T is raised the number of occupied vibrator levels increases proportionately, and the consequent increased number of transitions needed to reach equilibrium compensates for the greater speed. Nothing of this sort can of course occur with the two-level system.

The outcome of the argument in this section, so far as the energy is concerned, confirms Einstein's analysis in which the terms proportional to $\bar{n}$ in (60) represent stimulated processes while the additional $\mathscr{P}_2/\tau_e$ describes the spontaneous deexcitation. At the time of Einstein's work, which took place in the context of Bohr's quantum theory, virtually nothing could be said concerning such linear response functions as the dipole oscillations. It would be tempting to suppose that the impulse response would continue

79

oscillating undamped until a sudden transition, as envisaged by Bohr, stopped it dead. The random occurrence of such transitions would lead in the average, for an assembly of systems, to exponential decay, not with time-constant τ_p but with the same time-constant, τ_e or τ_e', as for the energy. One must therefore be careful, when using Einstein's model for visualization of the process, not to apply it blindly. Given care, however, it is a very real aid to thought and, of course, was historically of the greatest importance.

The frequency-dependent susceptibility

The Bloch equations (53), with their solution (55) for $\rho_0\,\mathrm{e}^{-i\omega t}$ when the applied field is $\mathscr{E}_0\,\mathrm{e}^{-i\omega t}$, yield a complex susceptibility $\chi(\omega)=p_0\rho_0/\mathscr{E}_0$ of characteristically resonant form for the case of equal wells. When the wells are unequal various modifications must be made to the Bloch equations, the details of which are left to the reader. One must bear a number of points in mind. It follows from (26) that in the representation of fig. 4(*a*) the equations of motion of the loss-free system are unchanged in form, though ω_0 is now $2E/\hbar$ and p_0 must be replaced by p_{12}, i.e. $p_0\Delta_0/E$, from (19). Proceeding to (53), we note that τ_p and τ_e (better, τ_p' and τ_e') must be multiplied by Δ_0/E, and that $\zeta_0=-\tanh(E/k_BT)$. This last point is significant, since ζ now suffers a first-order oscillation as a result of the dependence of E on $\mathscr{E}$; and since $\langle p\rangle$ is obtained by projecting onto a tilted plane, this oscillation of ζ makes a contribution to the susceptibility. This contribution is indeed the same as that attributable to the varying base-line in the step-response function, fig. 3(*b*). The reason for not displaying this approach to the problem in detail is that it is easier to incorporate the various modifications into the response function, and derive $\chi(\omega)$ therefrom. Thus (29), with f put equal to $\tanh(E/k_BT)$ and the oscillatory term appropriately damped, supplies the first term of (62). The second term rises smoothly from 0 to its final value $(\Delta^2p_0^2/E^2k_BT)\,\mathrm{sech}^2(E/k_BT)$ which is $-(\Delta p_0^2/E)\,\mathrm{d}\zeta_0/\mathrm{d}\Delta$; unit field changes the equilibrium value of ζ_0 by $p_0\,\mathrm{d}\zeta_0/\mathrm{d}\Delta$, whose projection onto the tilted plane gives a change in $\langle p\rangle$ of $-p_0\Delta/E$ times this. Hence

$$S(t)=(\Delta_0^2p_0^2/E^3)\tanh(E/k_BT)[1-\mathrm{e}^{-Et/\Delta_0\tau_p'}\cos(2Et/\hbar)] \\ +(\Delta^2p_0^2/E^2k_BT)\,\mathrm{sech}^2(E/k_BT)(1-\mathrm{e}^{-Et/\Delta_0\tau_e'}). \qquad (62)$$

The impulse-response is $\mathrm{d}S/\mathrm{d}t$:

$$h(t)\doteqdot(2\Delta_0^2p_0^2/\hbar E^2)\tanh(E/k_BT)\,\mathrm{e}^{-Et/\Delta\tau_p}\sin(2Et/\hbar) \\ +(\Delta^2p_0^2/E\Delta_0k_BT\tau_e)\,\mathrm{sech}^2(E/k_BT)\,\mathrm{e}^{-Et/\Delta_0\tau_e}, \qquad (63)$$

if $E\tau_p/\hbar\gg1$ and the oscillations are weakly damped. Finally, according to

(5.4) the susceptibility $\chi(\omega)$ is the Fourier transform of $h(t)$:

$$\chi(\omega) = \int_0^\infty h(t)\,\mathrm{e}^{\mathrm{i}\omega t}\,\mathrm{d}t$$
$$\doteqdot (4\Delta_0^2 p_0^2/\hbar^2 E)\tanh(E/k_\mathrm{B}T)/(\omega^2 - 4E^2/\hbar^2 + 2\mathrm{i}\omega E/\Delta_0\tau_p)$$
$$+ (\Delta^2 p_0^2/E^2 k_\mathrm{B}T)\,\mathrm{sech}^2(E/k_\mathrm{B}T)/(1 - \mathrm{i}\omega\Delta_0\tau_\mathrm{e}/E). \qquad (64)$$

The imaginary part of χ is shown in fig. 3(c), with two peaks corresponding to the two terms of (64). The first gives the peak on the right of the diagram, a typical Lorentzian resonance, centred on $2E/\hbar$, whose width at half power is $2E/\Delta_0\tau_p$; to the left of the diagram is the Debye-type relaxation peak[3] contributed by the second term. The peaks are comparable in width but very different in form, although this may not be immediately obvious in the conventional plot of χ'' against $\log\omega$, rather than against ω itself. Since $\chi'' = \chi'(0)\omega\tau/(1+\omega^2\tau^2)$, we derive the form of the logarithmic plot by writing λ for $\ln\omega$ and λ_0 for $\ln(1/\tau)$:

$$\chi''(\lambda) = \chi'(0)\,\mathrm{e}^{\lambda-\lambda_0}/(1+\mathrm{e}^{2(\lambda-\lambda_0)}) = \chi'(0)\,\mathrm{sech}(\lambda-\lambda_0), \qquad (65)$$

a symmetrical bell-shaped curve superficially resembling a Lorentzian resonance if we overlook its width – between the half-peak points there is a factor of 13.9 in frequency, the two half-points being defined by $\omega\tau = 2 \pm \sqrt{3}$. On the logarithmic plot of fig. 3(d) the relaxation peak has a width of 1.14, whatever the value of τ which only determines the position of the peak ($\omega\tau = 1$), while the resonant peak has a width of about $1/Q$, much less than the relaxation peak even for a low-Q resonance.

The area under each curve on the logarithmic plot, i.e. $\int_0^\infty \chi''\,\mathrm{d}\lambda$, is a convenient measure of the strength of the process. According to the

I.110 Kramers–Kronig relations (5.9) the area is $\frac{1}{2}\pi$ times $\chi'(0)$, the contribution of the process to the static polarizability. Hence, from (64), we have the two strengths which should be compared with the two terms in (23):

$$\text{Resonance strength} = (\pi p_0^2\Delta_0^2/2E^3)\tanh(E/k_\mathrm{B}T), \qquad (66)$$

$$\text{Relaxation strength} = (\pi p_0^2\Delta^2/2E^2 k_\mathrm{B}T)\,\mathrm{sech}^2(E/k_\mathrm{B}T). \qquad (67)$$

These strengths are independent of the relaxation times and are determined by the tunnelling probability, as measured by Δ_0, and the asymmetry Δ of the potential wells in relation to $k_\mathrm{B}T$. The resonance has its greatest strength of $\pi p_0^2/2\Delta_0$ at low temperatures and when $\Delta = 0$, but the relaxation strength is zero when Δ and $T = 0$, having a maximum value of $0.27p_0^2/\Delta_0$ when $\Delta = \sqrt{2}\Delta_0$ and $k_\mathrm{B}T = 2.24\Delta_0$. Under the same conditions the resonance strength is reduced to $0.20p_0^2/\Delta_0$, principally because of E^3 replacing Δ_0^3 in the denominator of (66). When the wells are markedly uneven relaxation can still proceed by the particle tunnelling from the higher well to the lower, but there is very little of an oscillatory dipole associated with this process. When the temperature is high enough for $\tanh(E/k_\mathrm{B}T)$ to be replaced by E/k_BT and $\mathrm{sech}^2\,(E/k_BT)$ by unity, the two expressions, (66) and (67), sum to $\pi p_0^2/2k_BT$, independent of Δ_0 or Δ. A static susceptibility varying inversely with temperature is just what is predicted by the simplest theories of orientable permanent dipoles, as in Curie's law for paramagnetic materials and Debye's analysis of dipolar dielectrics. We shall comment further on this point in due course, with specific reference to the ammonia molecule.

The relaxation peak that arises by quantum-mechanical tunnelling between potential wells is very much rarer than the loss peaks of the same form which are found in dielectrics,[3] magnetic materials,[4] elastic materials[5] and many others, where the exponential approach to equilibrium after application of a step-function perturbation is by thermal activation over a potential barrier. These processes are unaccompanied by any resonant behaviour and will be considered no further here, except to remark that, through the operation of the Boltzmann factor, the time-constant for crossing the barrier increases rapidly at low temperatures, usually as $\exp\,(T_0/T)$, where T_0 is a characteristic temperature commonly in the range around 10^4 K. With quantum-mechanical tunnelling, however, we may expect a much slower temperature variation. According to (60), τ_e' varies inversely as the mean energy (including zero-point energy) of the modes at frequency $2E/\hbar$ which are the agents of the relaxation process; because of spontaneous emission, τ_e' tends to a constant as T falls to zero, rather than rising indefinitely as it does when thermal excitation is responsible.

Dielectric loss in polyethylene – an example of tunnelling relaxation

Although pure polyethylene, consisting of very long chains of $—CH_2—$ units, is remarkably free from dielectric loss at low temperatures, when lightly oxidized by heating in air it develops a pattern of loss that conforms very well to the Debye model and can be explained in detail as arising from a sparse distribution of double-well potentials between which a proton can tunnel. The chemical configuration of these wells is a matter of uncertainty which is, however, irrelevant to our purpose. What emerges from the study of this material is that the two wells are reasonably similar in depth – though not exactly equal, for then they would show no relaxation peak. The experimental curves, fig. 6(a), of the frequency-variation of χ''/χ' exhibit peaks superimposed on a low background loss whose exact value is somewhat conjectural but certainly consistent with the observed loss in unoxidized material. After allowance for the background, the peaks conform to the Debye pattern (65) very well indeed; the highest peak, for example, has a width at half-peak amounting to a factor of 14.5 in frequency, only slightly larger than the theoretical 13.9. This could not happen if there were a substantial spread in relaxation times among the different double-well systems, and it may be inferred that they are all closely similar in

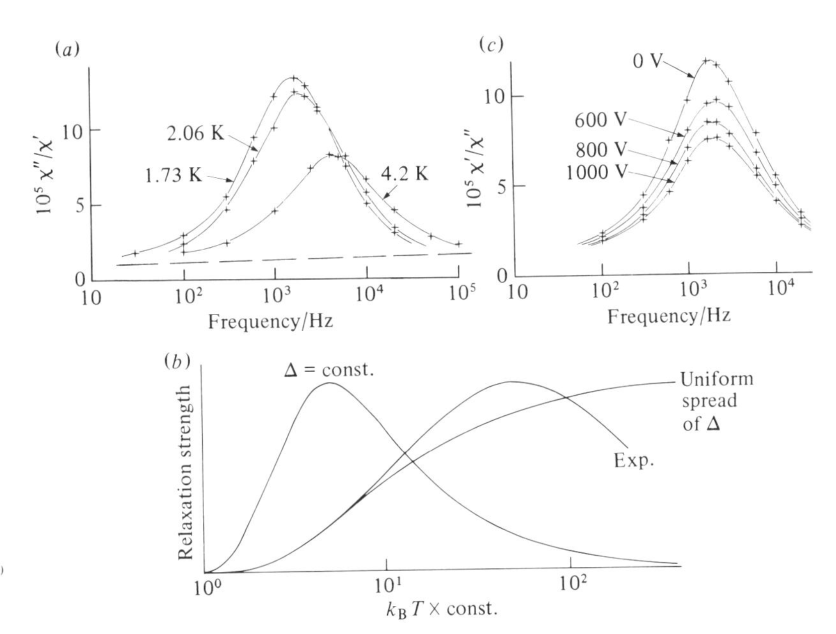

Fig. 6. (*a*) Dielectric relaxation loss in polyethylene (Phillips[2]); the temperatures at which the measurements were made are shown beside the curves. (*b*) Theoretical curves for the temperature-variation of relaxation strength for two models – an assembly of identical unequal double-wells (Δ = const) and an assembly having a uniform wide spread of Δ – and an experimental curve (Frassati and Gilchrist[6]). (*c*) Application of a steady strong electric field polarizes the double-wells and lowers the relaxation strength without affecting the relaxation frequency.[2]

structure and environment. Too much, however, must not be read into this; the very width of the Debye peak has the consequence that a moderate spread of relaxation times, and hence of peak positions, produces little extra breadth. A spread over a factor of 1.5 is not inconsistent with the observations.

The curves in fig. 6(*a*) show how the relaxation strength increases as the temperature is lowered, but this cannot continue down to zero temperature and the form of the variation is revealing. Let us suppose, as is fully justified by detailed experimental study, that the double-well systems differ from one another principally in their asymmetry. Ideally, Δ would be zero and the wells identical, but local strains and other environmental factors disturb the potential so that Δ may be taken to vary randomly over what turns out to be a relatively narrow range. It is easy to compute from (67) how the area under the relaxation loss curve will vary with temperature for a given choice of the spectrum of Δ, and two extreme examples are given in fig. 6(*b*), one for constant Δ and one for a uniform spread of Δ between $\pm\infty$; the curves are scaled to the same peak height, as are the experimental observations. The fall-off of the strength at higher temperatures is a sure indication that the spread of Δ is finite, though the behaviour at the lowest temperatures, which is dominated by the systems for which Δ is near zero, shows that a limited spread around zero is the model to be preferred over that of a single non-zero Δ. The temperature range in which the relaxation strength disappears as T is lowered is the most obvious indicator of Δ_0, while the fall-off at higher temperatures gives a good measure of the range of Δ. For this sample the disappearance at low temperatures is well under way at 0.1 K, and the entire variation with temperature can be accounted for by taking Δ_0/k_B to be 0.045 K (a level splitting of 7.8 μeV, corresponding to a resonant frequency for excitation to the upper level of 1.9 GHz), while the spread of Δ is over a range of about $\pm 70\Delta_0$, which is still very small in actual energy, $\pm\frac{1}{2}$ meV.

If the temperature is low enough so that $k_B T$ is distinctly less than the spread of Δ (which is achieved by working below the peak in fig. 6(*b*)) the effect of a steady field of moderate strength is to displace all Δ by $\boldsymbol{p}_0 \cdot \boldsymbol{\mathscr{E}}$. The susceptibility is not significantly changed, since the process merely replaces one set of polarizable systems, those for which $E \lessapprox k_B T$, by another and virtually equivalent set. A strong enough field, however, will eliminate all systems by ensuring that all Δ's are greater than $k_B T$. In practice the systems are randomly oriented, so that some lie nearly normal to $\boldsymbol{\mathscr{E}}$ and are not seriously affected. Nevertheless, the loss of relaxation strength on applying a steady field is clear from fig. 6(*c*), and the combination of this information with estimates of the spread of Δ derived from the temperature variation enables p_0 to be inferred, and hence the separation of the potential wells. Further, the resonant frequency is controlled by the tunnelling probability according to (13), so that the height of the potential barrier can now be estimated. All this makes clear that means are available to determine the parameters of the system experimentally; the outcome is a model potential which not only has the required properties to account for all the data quantitatively, but is also a very reasonable double-well of such dimensions and with such a barrier as might be expected for a proton (i.e. hydrogen ion) in a hydroperoxide group, —OOH, which is one conjectured source of double wells.

The frequency at which the relaxation peak occurs is determined by the strength of the coupling of the double-well to the thermally maintained vibrations of its environment, for it is these that supply or remove the energy involved in the transitions between levels that restore and maintain the state of equilibrium. With the low frequency (1.9 GHz) and modest dipole moment (6×10^{-30} Cm) the time constant for equilibration by spontaneous electromagnetic radiation is far too long (~40 years) to be significant. We know, however, from the spread of Δ among otherwise identical systems that the potential is influenced by strain, as if there were a polarizing electric field associated with the deformed lattice. The double-well is thus coupled to the lattice vibrations (phonons) of the host polymer, and the tunnelling process is in consequence referred to as phonon-assisted tunnelling. For non-dispersive waves such as lattice waves or light the density of states g at a given frequency is inversely proportional to the cube of the wave velocity v_s, which in polyethylene is about 2×10^5 times less than c. If a single elastic mode were to generate the same Δ as an electromagnetic mode with the same energy density, the two would interact identically with the double-well system, and the change in g would reduce the time constant by $(v_s/c)^3$, from 40 years to 2×10^{-7} s. In fact the coupling to the lattice strain must be markedly less, not only because the time constant at the lowest temperatures (where spontaneous processes dominate) is 10^3 times larger than this, but because it would be hard to understand otherwise the limited spread of Δ. As a result of anisotropic contraction on cooling, strains of the order of 1% must exist in the polymer, giving rise to strain energy densities of about 10^5 J m^{-3}. An electric field having this energy density, $\mathscr{E} = 1.5 \times 10^8$ V m^{-1}, would generate a value of Δ equal to 6 meV, more than 10 times the observed spread. Since the time-constant varies inversely as the square of the coupling constant, there is at least a factor of 100 here towards the 1000 needed. In view of the rough arguments involved, this is near enough to suggest that a consistent picture is emerging.

One last point is worthy of notice. In view of the spread of E, and hence the resonance frequency, from Δ_0 to $70\Delta_0$, the very narrow range of relaxation times needs explanation. This in fact comes about automatically for the suggested coupling process. We have seen that τ_e for the equal-well system is reduced to $\tau_e\Delta_0/E$ by the introduction of inequality, and that the relaxation time in the presence of thermal radiation (phonons in this case) is reduced by a further factor $1/(2\bar{n}+1)$. Hence we expect τ to have the form $\tau_e\Delta_0/E(2\bar{n}+1)$. Since $2\bar{n}E$ is just k_BT for oscillators of level spacing $2E$, $\tau \sim \tau_e\Delta_0/k_BT$; Δ has vanished, and τ takes the same value for all.

Other examples of double-well systems; ammonia

The hydrogen molecule-ion, H_2^+, with one electron moving in the field of two protons, is as simple an example as one could hope to find, and it was the subject of many detailed calculations in the early days of quantum mechanics. The energy can be lowered by sharing the electron between the protons rather than having it bound to one only and, as expected, the electron wave-function in the ground state is symmetric and describes the bound state of the ion. In this case the barrier is very transparent and in the equilibrium configuration the minimum of $\psi\psi^*$ between the protons is shallow.

A considerable number of molecules[7] show a variety of stable configurations and can transform from one to another by tunnelling or by thermal excitation over the barrier. In ethane $H_3C{\cdot}CH_3$ the two configurations fig. 7(a) and (b) differ in energy by about 1/8 eV, corresponding to a temperature of 1500 K, and the CH_3 groups can rotate through 120° from one stable position of the form (b) to another by passing over or tunnelling through this potential hump. The tunnelling frequency (~10 GHz) is about 100 times lower than the frequency of torsional oscillations around a stable position, so that the ground state is slightly split, but not by enough to make a great difference to the bonding energy. In dimethyl-acetylene $H_3C{\cdot}C \vdots C{\cdot}CH_3$ the CH_3 groups are further apart and the potential hump much smaller; in this case a good approximation to the energy results from assuming that the groups rotate freely, with a hardly significant correction for the correlation of their angular positions.

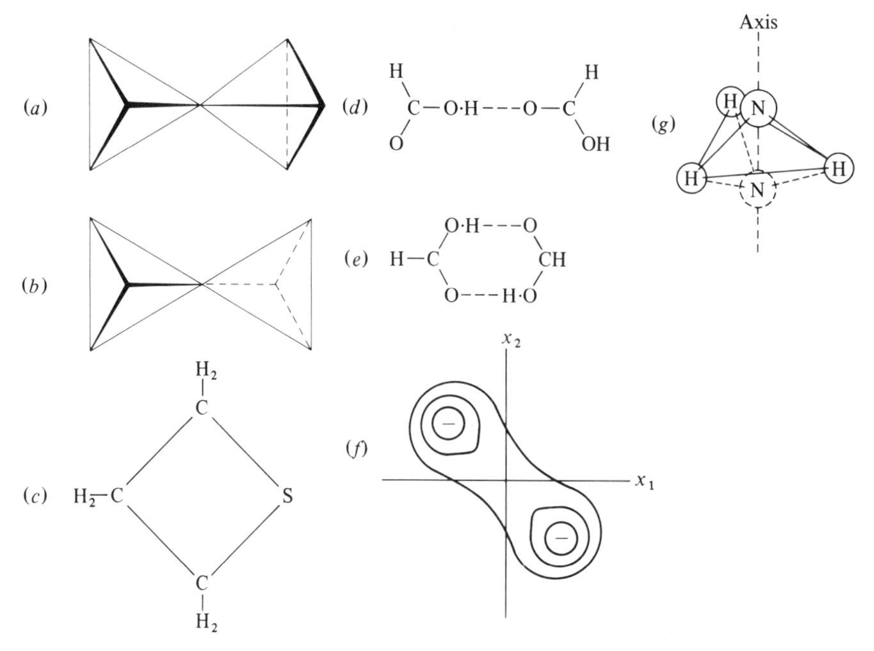

Fig. 7. (*a*) and (*b*) two configurations of ethane $(CH_3)_2$, shown as tetrahedra with carbon atoms at their centres, single-bonded at points in contact, and having H-atoms at the free points. In (*a*), the cis-configuration, the H_3 triangles at either end are similarly oriented; in (*b*), the trans-configuration of lower energy, they are 60° apart. (*c*) Trimethylene sulphide; the potential minima for the S-atom lie just above and just below the plane of the paper. (*d*) Two formic acid molecules bonded by a single hydrogen bond. (*e*) Two formic acid molecules bonded by two hydrogen bonds. (*f*) Schematic representation of the potential energy surface for (*e*). (*g*) Tetrahedral structure of ammonia, showing the two equivalent energy minima for the N-atom relative to the H_3-triangle.

The hindered rotation of CH_3 groups such as those, or of a polymer chain such as $(CH_2)_n$ about any one of the C—C bonds, makes possible a large variety of stable molecular configurations differing rather little in energy and separated by potential barriers. Important though these are to the polymer chemist, they are too complicated for our present purpose, as well as showing very little in the way of tunnelling; to switch from one configuration to another involves moving too many massive atoms too far, and is normally accomplished only by thermal excitation. On the other hand ammonia and trimethylene sulphide exhibit readily discernible tunnelling behaviour in the splitting of their ground states by the coupling of two alternative equivalent configurations. In view of its importance in the development of the maser, to which we shall turn in chapter 20, we shall pay special attention to ammonia, and in a few words dispose of trimethylene sulphide. The ring of 3 carbon atoms and 1 sulphur (fig. 7*c*) has its lowest energy when bent (*puckered*), and we may imagine the carbon atoms held fixed while the sulphur moves normal to the diagram in a symmetrical double well with the potential maximum lying in the plane of the carbons. This exaggerates the difficulty experienced by the sulphur in tunnelling from one side to the other; when the molecule is free the carbons also move and the sulphur behaves as if its mass were considerably less than its real mass. We shall not derive the required correction, which is exactly analogous to the reduced mass correction required for nuclear motion in the theory of the energy levels of hydrogen-like atoms, and is straightforward to calculate. Since the molecule has a component of dipole moment normal to the carbon plane, which reverses after the sulphur changes sides, the tunnelling process carries an oscillatory dipole moment and transitions between the symmetrical ground state and the antisymmetrical state, 34 μeV above, can be detected as an absorption line in the microwave spectrum at 8.2 GHz. By contrast, when the sulphur atom is replaced by oxygen the barrier is so low that the ground state lies above it; its presence is, howcvcr, rcadily apparent in the microwave spectrum around 24 GHz since the vibrations of the oxygen atom are highly anharmonic.

One more example is worth mention, though the experimental evidence is not so clear. Carboxylic acids tend to form dimers, of which the dimer of formic acid HCOOH is the simplest example. If the molecules were brought together as in fig. 7*d* the hydrogen bond would have the proton much closer to its original carbon atom than to the other, and it would be sitting in a simple potential well. Electrostatic repulsion prevents it transferring to the other molecule, but this difficulty vanishes when two hydrogen bonds are present, as in fig. 7*e*. There are two equivalent configurations, mirror images, and it is probable that they, rather than a configuration with both protons placed midway, are the states of lowest potential energy. This provides a model of a tunnelling process in which two atoms tunnel simultaneously, and serves to illustrate what may be even more complex in other cases, with more than two atoms tunnelling and more than one route connecting the stable states. A simplified representation is probably adequate to illustrate the principal points. Since the timescale for proton tunnelling is slow in comparison with other vibrations and with electron rearrangements, the Born–Oppenheimer (adiabatic) approximation[8] allows us to imagine that the rest of the molecule has plenty of time to adjust itself to changes of the probability distribution of the protons; they are thus to be thought of as moving in a potential which is a well-defined function of their position co-ordinates. And we may further simplify the problem by making the proton motion one-dimensional, along the line joining its own oxygen atoms. In this way the problem is reduced to motion of a particle of protonic mass on a plane in which the potential is $V(x_1, x_2)$, x_1 and x_2 being the co-ordinates of the two protons. The general form of $V(x_1, x_2)$ is easily guessed – there must be minima at points corresponding to the stable configurations, e.g. fig. 7*e*, and nowhere else; any departure from the line $x_1+x_2=0$ means a different net charge on the two molecules, with extra electrostatic energy, so that the central position, $x_1=x_2=0$, is probably a saddlepoint of V, as shown in fig. 7*f*. The simultaneous tunnelling of the two protons is now described by a simple two-well model, the wells being further apart by a factor $\sqrt{2}$ than the stable positions of the individual protons; this may well reduce the tunnelling probability by a large factor, the same as would arise if we kept the original well separation and doubled the mass of the tunnelling particles. The ground state energy is lowered a little by the tunnelling, and corresponds to a symmetrical wave-function, with the antisymmetrical state slightly above. In both states $\psi\psi^*$ has equal maxima in the two wells, so that one may say that *if* one proton is found to be displaced to the right, there is a high probability that the other will be displaced to the left; and *vice versa*. Both configurations are equally likely. In a mixture of the two states the configurations interchange periodically, but there is no oscillatory dipole associated with the process, only a quadrupole which will be very weakly coupled to an electromagnetic field and correspondingly difficult to detect.

Let us now turn to the especially important case of ammonia, to which is traditionally ascribed the tetrahedral structure of fig. 7*g*. The centroid of negative charge is not compelled by symmetry to coincide with the centroid of positive charge, and one must expect to find a dipole moment oriented along the axis of the molecule. In fact the dipole moment is measured to be 1.48 Debye units, or 4.94×10^{-30} Cm, equivalent to one electron displaced through 0.03 nm (about half the Bohr radius). This model is, however, too classical since an equivalent structure is obtained by taking the nitrogen atom through the hydrogen plane to a mirror position opposite, and there is ample evidence that this inversion proceeds quite readily by tunnelling. In the spirit of the last paragraph we may suppose the only co-ordinate of significance to be the position of the nitrogen atom on a line normal to the hydrogen triangle, passing through its centroid. If a nitrogen atom has mass m_N and a hydrogen atom mass m_H, the effective mass of the tunnelling particle is m^*, where $1/m^* = 1/m_N + 1/3m_H$, i.e. $m^* = 2.5$ hydrogen masses. The chemical evidence which makes the tetrahedral structure so natural must now be interpreted as evidence for two equivalent potential wells separated by a barrier. The ground state is described by a symmetrical wave-function, with the antisymmetric state lying a little above it, and although neither state in itself has a dipole moment the superposition of the two produces an oscillatory moment exactly along the lines already discussed. The inversion phenomenon had been inferred from the line-splitting in the infra-red spectrum of ammonia, and extensively studied theoretically, before it was observed in the plainest manner as an absorption line in the microwave region, about 24 GHz, resulting from the direct transition between the split levels. This was in 1934, with the most primitive of equipment, and it gave microwave spectroscopy its first success. After 1945, with better equipment, the absorption at lower pressures could be studied, and it then became clear that there were many lines in the spectrum; more than 60 have been detected, of which about half are to be seen in fig. 8. They arise because the many rotational states of the molecule that are excited at room temperature suffer different centrifugal distortions of the bond lengths. In consequence the potential barrier and the level splitting are different in each rotational state. In addition, the nitrogen nucleus has a quadrupole moment that interacts with the field of the nitrogen atoms, and there are other internal effects as well. These complications fortunately need not be allowed to interfere with the essential simplicity of the following discussion, but we shall return to them briefly in the next chapter. **126**

There would seem at first sight to be a contradiction between the idea of a permanent dipole, as measured, and the symmetry of the charge

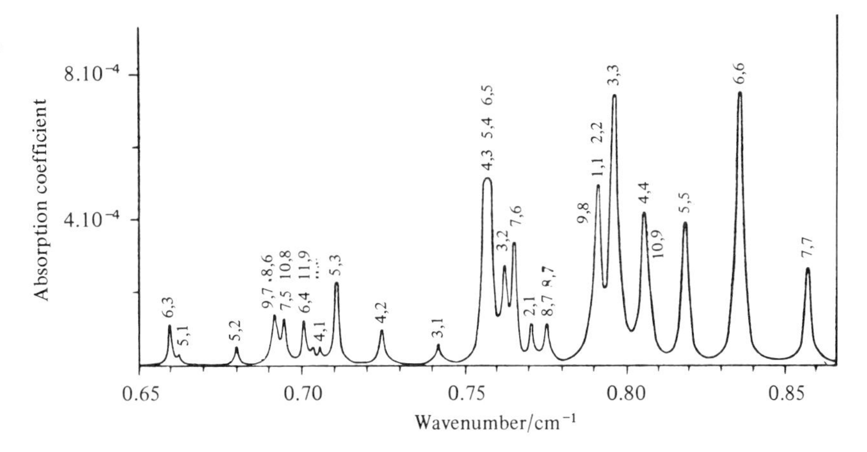

Fig. 8. Microwave absorption in ammonia gas at room temperature and a pressure of 1.2 Torr (Bleaney and Penrose[9]).

distribution in the stationary states, but this is resolved by a calculation which shows that the polarization induced by an electric field in the initially unpolarized stationary states simulates the polarization by orientation of permanent dipoles, provided $k_B T$ is greater than the level splitting. The necessary calculation has already been carried out for the double-well system; according to (23), if the wells are equal $\bar{\alpha} = (p_0^2/\Delta_0)\tanh(\Delta_0/k_B T) \approx p_0^2/k_B T$ if $k_B T \gg \Delta_0$. And this is just what is found for a permanent dipole which is allowed two orientations, either along or against the field. The two states will be occupied in the proportion $e^{\pm p_0 \mathscr{E}/k_B T}$, or $1 \pm p_0\mathscr{E}/k_B T$ when $\mathscr{E}$ is small; the excess fraction, $p_0\mathscr{E}/k_B T$, along the field gives a mean polarizability of $p_0^2/k_B T$. In the case of ammonia, with $\Delta_0/k_B \sim 0.6$ K, both models should yield virtually the same value of $\bar{\alpha}$ at all temperatures above 2 K.

This example illustrates the irrelevance, in many practical applications, of an apparently basic distinction between the classical and quantal views of symmetry. The wave-function in the ground state of ammonia is typical of many ground states which reflect the symmetry of the potential. If the potential is a symmetrical double well, $\psi\psi^*$ is symmetrical in all stationary states, but classically the particle must rest in either well, not simultaneously in both. With a potential barrier so high that it would take a year for tunnelling to occur it would be pedantic to insist on interpreting the observation of the particle in one well in terms of the superposition of two states differing in energy by 10^{-41} J, unless one had means of separating the two states, and sufficient freedom from thermal disturbances for a time exceeding the tunnelling time. In this case the energy splitting corresponds to a temperature of 10^{-18} K – only in fantasy could one pretend that the 'correct' symmetry-maintaining quantal picture has any advantage over the symmetry-breaking classical picture. By the same argument, so long as we are concerned with the behaviour of ammonia molecules in reasonably close contact with an environment well above 2 K, the classical picture is adequate and easier to use than the quantal. But now it is not difficult to isolate them for much longer than their tunnelling time of 10^{-11} s; at their low inversion frequency radiative coupling to the outside world is extremely weak, and at only a moderately low pressure the mean time between collisions is greater than this. In the ammonia maser[(10)] a microwave cavity is excited into oscillation by molecules in the antisymmetric state which pass through and are stimulated to make transitions to the ground state. The operation thus relies upon the possibility of separating antisymmetric from symmetric states and on keeping the molecules apart for long enough to make use of the specially prepared population of antisymmetric states. It is now imperative to abandon the classical picture and think of the quantal description.

The ammonia maser will be a central feature of chapter 20, but since the separation process is so closely related to the discussion of the double-well system it will be convenient to describe it here. Separation was achieved in the original maser by an electrostatic lens, making use of the positive polarizability of the symmetric state and the equal, but negative, polarizability of the antisymmetric, $\alpha = \pm p_0^2/\Delta_0$ from (21). In an inhomogeneous electric field the symmetric state can lower its energy still further by moving towards a region of stronger field, and is attracted in this direction, while the antisymmetric state has its energy raised and is therefore repelled. A field of magnitude $\mathscr{E}$ induces a dipole moment $\alpha\mathscr{E}$ and the energy is thereby reduced by $\frac{1}{2}\alpha\mathscr{E}^2$; consequently the force is grad $(\frac{1}{2}\alpha\mathscr{E}^2)$. By means of an array of four cylindrical electrodes with the cross-sectional form shown in fig. 9, a transverse electric field of quadrupole symmetry is established.

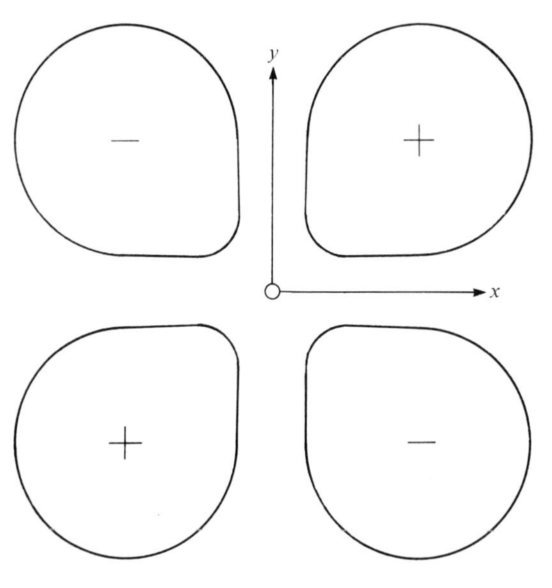

Fig. 9. Cross-section of quadrupole electrostatic lens used to focus ammonia molecules in the antisymmetric state (Gordon, Zeiger & Townes[(10)]).

The inner cheeks of the electrodes have the form of rectangular hyperbolas, xy = constant, compatible with the desired potential near the axis,

$$V = axy,$$

a being a constant. The field $\mathscr{E}$ is $-\text{grad } V$, i.e. $-a(y, x)$ and $\mathscr{E}^2$ is just $a^2(x^2+y^2)$, i.e. a^2r^2. The force on a molecule is therefore radial, drawing the antisymmetric molecules towards the axis and repelling the symmetric. If an axial but slightly divergent stream of atoms emerges from a source into the evacuated space of the lens, only atoms in the upper state are focussed back towards the axis and, passing through a hole in a diaphragm, are rather thoroughly separated from the atoms in the ground state. To appreciate the technical problem let us calculate the distance between the source and the point of focus. The equation of radial motion of a molecule is the equation of simple harmonic motion, $m\ddot{r} = -\text{grad}\,(\frac{1}{2}p_0^2a^2r^2/\Delta_0) = -p_0^2a^2r/\Delta_0$, and the period of half a cycle is $(\pi/p_0a)(m\Delta_0)^{\frac{1}{2}}$. A molecule moving along the axis with velocity v_z travels $(\pi/p_0a)(2E\Delta_0)^{\frac{1}{2}}$ in this time, where E is $\frac{1}{2}mv_z^2$, the molecule's kinetic energy, about $\frac{1}{2}k_BT$. In the original experiment a took the value 6×10^8 V m^{-2} and the ammonia molecules issued from a reservoir at room temperature. If we use for p_0 the measured dipole moment and for $2\Delta_0/\hbar$ the inversion frequency, 24 GHz, the focussing distance is calculated to be about 20 cm, much the same as was found in practice. The value of a deserves comment – 1 cm from the axis the field strength would be 6×10^6 V m^{-1}, so that rather high potentials are needed on the electrodes, but not so high as to ruin the experiment by arcing.

It should be noted that the Maxwellian distribution of velocities gives different molecules different focussing distances, so that there is a measure of velocity selection. Nevertheless the velocities of those that actually enter the cavity, and hence the time they take to pass through, are liable to spread over a rather wide range; moreover, without detailed knowledge of the geometry of the arrangement, including the sizes of orifices, no estimate can be made of this distribution. This places certain restrictions on comparisons between theory and experimental performance of a maser, on the basis of published information, but it will become clear in chapter 20 that more interest attaches to the general physical principles than to details of behaviour.

19 Line broadening

The natural line broadening resulting from electromagnetic or acoustic radiative processes, the only causes of broadening discussed so far, by no means exhausts the mechanisms available and indeed is usually so minor an effect as to be of small practical importance. We may distinguish broadening due to different behaviour on the part of different members of an ensemble from broadening exhibited by each member on its own. In the first class are Doppler broadening and broadening due to variations in environment, not to mention the effects of slight differences between different, superficially similar, systems (e.g. different isotopes). In the second class, in addition to radiative processes, must be counted anything, especially collision with other atoms, which interrupts or distorts the wave-train emitted by a single system so as to widen its spread of Fourier components. There is a very large literature(1) on these effects, whose detailed analysis is both taxing and controversial. No attempt will be made here to go beyond an elementary discussion and illustration of some of the leading ideas, with examples of how the line-width may be reduced or its effects mitigated for the purpose of high-precision measurements of the central frequency. We start with the second class of processes, and for our purpose the two-level system provides an adequate model, with ammonia as a practical realization.

69

The absorption spectrum of ammonia at a rather low pressure, 1.2 torr, in the wavelength range from 1.1 to 1.5 cm is shown in fig. 18.8, each line resulting from transitions between pairs of levels in different rotational states of the molecule, as defined by the pairs of quantum numbers above each line. The first, J, may be visualized as defining the total angular momentum, while K defines that component of J along the axis, normal to the plane of hydrogen atoms, as indicated in the vector diagram of fig. 1. When $K = J$ the molecule is essentially spinning about this axis, and the hydrogens are thrown our centrifugally, to lower the potential barrier and increase the splitting of the pair of levels. The magnitude of the effect is determined roughly by J^2, and this explains the systematic progression from 1,1 to 7,7 at the right of fig. 18.8. Where K is less than J there is also a component of angular momentum about an axis orthogonal to the former axis, and the centrifugal effect of this elongates the barrier, so that lines with large J and small K appear at the left, low-frequency, side. The widths of the lines are real, not instrumental, and orders of magnitude greater than can be explained by radiative broadening or the Doppler effect. They are proportional to gas pressure; at 100 torr, 80 times higher than was used for fig. 18.8, the lines are so broadened as to overlap, and the resulting absorption spectrum shows no fine detail.

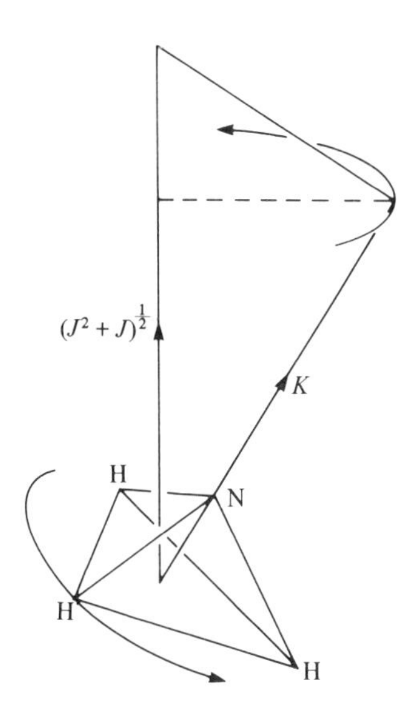

Fig. 1. Vector diagram of angular momentum of the ammonia molecule. The total angular momentum, $(J^2+J)^{\frac{1}{2}}$ units of $\hbar$, is drawn vertical, and $\boldsymbol{K}$ represents the component along the axis of the molecule, whose instantaneous position is drawn at the foot of the diagram; $\boldsymbol{K}$ precesses about the axis of total angular momentum.

The line-broadening is not usually so marked when the experimental chamber contains only a small amount of ammonia and the pressure is made up with a non-polar gas. Thus to obtain the same broadening in a mixture which is predominantly helium, with a trace of ammonia, 16 times the pressure is needed as in pure ammonia. The magnitude of the broadening observed with helium is close to what would be expected if every impact of a helium atom on an ammonia molecule disturbed the latter to such effect that its subsequent vibration was uncorrelated in phase to what had been taking place before. We may suppose the step-function response of a single molecule to proceed as in fig. 18.3(*a*) with negligible radiative decay, until a collision introduces an arbitrary phase shift; averaging over an ensemble of such molecules, all suffering their first collision at the same instant, we shall find that the wave-train stops abruptly at this point. But, of course, different molecules are hit at different times and the impacts are random, so that the probability of the wave-train lasting beyond a given time t decays exponentially, as e^{-t/τ_c}, where τ_c is the mean time between impacts of helium atoms on a given ammonia molecule. The theory is exactly analogous to the theory of the free path distribution in a gas. With an exponentially decaying response function the line shape, being the Fourier transform, is Lorentzian about the mean frequency, ω_1: $I(\omega) \propto [1+(\omega-\omega_1)^2\tau_c^2]^{-1}$. The line-width at half the peak intensity is $2/\tau_c$, and is proportional to the collision rate, i.e. to the pressure, as observed.

I.76

Phase diffusion

The description just given of line-broadening by collisions is perhaps a little cavalier in the sharp distinction made between catastrophic processes, which randomize the phase, and others which are assumed to be negligible. In reality one must expect a continuous range of phase disturbances, from a multitude of small changes arising from rather distant encounters to the truly catastrophic effects of a head-on collision. Let us examine the matter in a little more detail, taking as a model for an encounter some influence (e.g. the field of the intrusive molecule) that causes the representative point in fig. 18.4 to move during the duration of the encounter in some direction other than in its undisturbed motion round the ζ-axis. If the disturbance acts in any way like an applied electric field, we have already seen how, as in 18.4(b) this causes the axis of the orbit to tilt. An encounter that is completed during a fraction of a cycle at the natural frequency $2\Delta_0/\hbar$ will be effective only if the field is so strong as to change the frequency substantially, and this automatically tilts the orbit through a large angle. According to the phase in the cycle at the moment of collision, the point will be displaced upwards, downwards or sideways. Thus an assembly of molecules all originally represented by the same point on the sphere will soon appear as a spreading cloud of points; not only is there randomization of phase but also of ζ-co-ordinate, and both T_1 and T_2, the longitudinal and transverse relaxation times, are shortened by such sharp collisions. This is what must be expected in ammonia if the colliding molecule approaches as near as 1 nm. With a typical molecular velocity at room temperature of 700 m s^{-1}, the most effective period of the encounter lasts for something like 2 ps, during which the oscillation achieves only $\frac{1}{20}$ cycle; closer collisions have their main effect even more quickly. On the other hand, in optical transitions at 10^4 times the frequency, there is time for 500 cycles during the collision; a very small change of frequency, with the orbit correspondingly inclined at a very slight angle, can then seriously alter the phase, with hardly any perceptible shift in the ζ-direction. In such collisions T_2 is shortened but T_1 is hardly affected.

As the cloud of representative points is thus caused to diffuse over the sphere, its centroid moves towards the centre of the sphere and correspondingly the mean oscillatory dipole strength per molecule decreases. We shall now show that the decrease is exponential, not simply for catastrophic collisions but for any mixture of weak and strong collisions. For simplicity the argument will be confined to the situation more appropriate to optical lines, when the spread of points is around a line of constant ζ, but it can readily be generalized to include diffusion over the whole surface of the sphere. The only assumption needed is that the position of a point on this line is immaterial to the scattering process: if $P(\phi, \phi')\,\mathrm{d}\phi'$ is the chance that in a certain time interval a point at ϕ will be scattered into the range $\mathrm{d}\phi'$ at ϕ', we assume P has the form $P(\phi - \phi')$, and no explicit dependence on ϕ alone. This is a natural assumption when the collision lasts for so many cycles that the precise moment at which it begins is irrelevant. We do not assume that $P(\phi-\phi')=P(\phi'-\phi)$; indeed, since any electric field increases, never decreases, the orbiting frequency it would be mistaken to suppose forward and backward phase shifts to be equally likely. We now express any distribution of points, having density $A(\phi)$ on the line of constant ζ as a Fourier sum:

$$A(\phi)=\sum_{-\infty}^{\infty} a_n \,\mathrm{e}^{in\phi}, \text{ with } a_{-n}=a_n^* \text{ since } A \text{ is real.} \quad (1)$$

Consider how any one term in this sum is affected by scattering; a density that began as $a_n\,\mathrm{e}^{in\phi}$ has become $a'_n\,\mathrm{e}^{in\phi}$, where

$$a'_n\,\mathrm{e}^{in\phi'}=\int_0^{2\pi} a_n\,\mathrm{e}^{in\phi}P(\phi-\phi')\,\mathrm{d}\phi, \quad (2)$$

and if $P(\phi-\phi')$ is also expressed as a Fourier sum $\Sigma_{-\infty}^{\infty}\,p_m\,\mathrm{e}^{im(\phi-\phi')}$, only one term survives integration:

$$a'_n=2\pi p_{-n}a_n. \quad (3)$$

The form of the nth component is unchanged, but its amplitude is multiplied by $2\pi p_{-n}$, and since subsequent intervals produce the same relative effect, a_n simply varies exponentially: $a_n \propto \mathrm{e}^{-\alpha_n t}$, in which α_n may be complex and different for every Fourier component of $A(\phi)$. This does not, however, prevent the centroid from relaxing exponentially, since its position is determined by one component only. Thus

$$\bar{\rho}=\bar{\xi}+\mathrm{i}\bar{\eta}=\int_0^{2x} A(\phi)\,\mathrm{e}^{\mathrm{i}\phi}\,\mathrm{d}\phi\bigg/\int_0^{2\pi} A(\phi)\,\mathrm{d}\phi=(a_{-1}/\bar{A})\,\mathrm{e}^{-\alpha_{-1}t},$$

the appropriate exponential decrement having been supplied in the last expression. The imaginary part of α_{-1} modifies the frequency of the centroid as it spins round the ζ-axis (that is to say, the centre of the resonance line is shifted by α''_{-1}) while the Lorentzian broadening is determined by the real part of α_{-1}.

It must be emphasized that the model employed here, of a two-level system perturbed by a transient electric field, is altogether inadequate to describe the complexities of collisions between real molecules. The purpose of this section has been to indicate by a specific example how the exponential character of the response functions, and in consequence the Lorentzian form of the broadened line, are not accidents of a special collision mechanism, but rather to be expected as the norm. It is therefore permissible to replace the continuous range of interactions by an equivalent model in which all close collisions are treated as catastrophic and the rest ignored. The precise definition of a close collision must depend on the spectrum of phase changes, especially if diffusion on the sphere is mainly accomplished in small steps, but for rough estimating one will not usually go far wrong in assuming that scattering of the representative point through more than 1 radian is catastrophic. When the pressure is high, however, so that the spectral lines are considerably broadened and shifted, they begin to suffer shape changes as well, for which this crude modelling is inadequate and which can only be understood by a very thorough analysis of the collision mechanism.

The rather general result that diffusive spread of the representative points over the sphere leads to exponential decay of the oscillatory dipole prompts the question whether the exponential radiative decay discussed in the last chapter may not be described analogously as a progressive loss of phase information when an excited system is coupled into a cavity. Can one, in other words, retain the representation appropriate to an isolated system, or one subject to a determinate imposed force, when the perturbation results from interaction with other quantized systems? The answer has already been given – no; once two systems have been coupled their wave-functions can only be separated again by making observations, and for the purpose of predicting the probability of a given observation the combined wave-function must be retained.† The last paragraphs may legitimately be criticized on this score, but probably no great error has been incurred in what was only intended as a very rough analysis. We shall be obliged in the next section to take a little more care, but not to be so distrustful of elementary procedures as to throw away the classical models which have so far proved very helpful. As we have seen with the Bloch equations, the original representation involving motion on the surface of a sphere needed extension, when cavity modes were coupled to the double well, so that the representative point could leave the surface. In a somewhat analogous fashion we shall find in the next chapter that when a harmonic oscillator (a resonant cavity) is coupled to a host of other quantized systems (ammonia molecules) the equivalent classical ensemble of chapter 13 does not lose its usefulness, but simply needs to be extended to take account of new circumstances.

108

163

21

This does not mean that we may not use the idea of diffusion on a sphere as a convenience to describe the evolution of the expectation value of the dipole moment or the energy, and in this way put the dissipative mechanisms of radiation and collisions on an equivalent footing. One must always remember, however, that the ζ-co-ordinate of the representative point defines the *mean* value of p as a result of many identical trials; in any one trial the particle will be found in one well or the other, never anywhere else, and $\pm p_0$ are the only possible outcomes, with probabilities $\frac{1}{2}(1\pm\xi)$. In the same way a statistical distribution of points representing an assembly of identical systems is only significant as an aid to calculating the probability of finding a given number of systems with moment $+p_0$ and the rest with $-p_0$. As the assembly increases in size the probability distribution becomes sharply peaked; if the centroid of N points is at $\bar{\xi}$; $\frac{1}{2}N(1+\bar{\xi})$ systems will

† It is the instinctive reluctance to accept this proposition, especially as applied to systems which were once coupled but have since been mutually isolated, that has generated most of the paradoxes, such as that of Einstein, Rosen and Podolsky,[2] by which the foundations of quantum mechanics have been (unsuccessfully) assailed.

be found in the right-hand well, with an RMS deviation from this value of $\frac{1}{2}[N(1-\bar{\xi}^2)]^{\frac{1}{2}}$. For most purposes, when macroscopic assemblies are being considered, deviations from the mean are negligible, and only the position of the centroid is relevant to the observations.

Stark broadening and resonance broadening

When a spectral line is observed in the presence of a gas of polar molecules, the broadening tends to be rather larger than with non-polar molecules on account of the electric field of the dipoles, which can cause significant phase shifts even when the approach is not close enough to be regarded as a direct collision. The Stark effect,[3] the shift of a spectral line by a uniform strong electric field, was first detected by the use of fields of about 10^7 V m^{-1}, and a dipole of 1 Debye unit produces this order of field strength at a distance of 1 nm, several times a typical atomic radius. A free electron or proton produces an equal field from ten times as far away, and line-broadening effects in plasmas present a real problem in the use of spectral analysis for diagnosing the molecular processes at work. We shall not attempt to discuss this very difficult field and shall even be content to pass lightly over the slightly less difficult Stark broadening by polar molecules. To obtain a rather more quantitative measure of its magnitude, consider a dipole p passing at velocity v, with its closest distance of approach equal to r, so that its field, say $p/4\pi\varepsilon_0 r^3$, may be supposed to act for a time of order r/v. Let us see how close r must be for the resulting phase shift of the resonance line $(1s \rightarrow 2p)$ of a hydrogen atom to be 1 radian. The change of frequency, $\Delta\omega$, of this line in a field $\mathscr{E}$ is $3\hbar\varepsilon_0\mathscr{E}/2\pi me$, so that the critical distance is $(3\hbar p/8\pi^2 mev)^{\frac{1}{2}}$, i.e. about $\frac{1}{2}$ nm if $p = 1.5$ Debye units and $v = 700$ m s^{-1}. This is a rather stronger effect than we should have calculated by use of the double-well model, for which at this sort of field strength the Stark effect is quadratic in $\mathscr{E}$ and rather weak. The Stark effect is, to tell the truth, a good deal too complicated for simple models to be satisfactory, and we shall leave it at that in order to return to a line broadening effect which may be usefully discussed in terms of the double-well model. This is *resonance broadening*, resulting from the interaction between two identical, or nearly identical, molecules, and it can be very strong.

87

Let us first look at a completely classical model, with the molecules replaced by identical harmonic oscillators, and the pressure assumed low enough for only binary collisions to matter. Suppose that a step function electric field has set them all vibrating parallel to $\boldsymbol{\mathscr{E}}$ with the same amplitude and phase; the measured response function $S(t)$ is the sum of all individual responses. When any two approach closely enough to be appreciably coupled by their dipole fields, the normal mode co-ordinates are the sum and difference of the individual displacements, and these continue to vibrate with unchanged amplitude. In particular the sum, which is the measured quantity, is unchanged in amplitude. The phase of the vibration is, however, changed since the normal mode frequency is altered by the coupling, and if this frequency change is $\Delta\omega$ (i.e. if $2\Delta\omega$ is the frequency difference between the two normal modes) the phase shift for an encounter of duration t_0 is $t_0\Delta\omega$. We take $1/\Delta\omega$ as a measure of the length of encounter needed for scattering through 1 radian, i.e. for the contributions of the molecules concerned to be effectively removed from the response; an effective collision cross-section may be defined in terms of the distance of approach required to produce this phase change. A molecule with an optical transition of f-value unity may be modelled by an electron undergoing harmonic oscillations. When the displacements of the electrons in two such molecules are x_1 and x_2, the potential energy of coupling is of the order of $e^2x_1x_2/4\pi\varepsilon_0 r^3$ and this leads to normal modes displaced by $\pm e^2/8\pi\varepsilon_0 mr^3\omega_0$ from the mean, ω_0. An assumed encounter time of r/v leads to a phase shift of 1 radian if r is $e/(8\pi\varepsilon_0 m\omega_0 v)^{\frac{1}{2}}$, about 6 nm for sodium atoms at a temperature of 1000 K. If each atom presents a target area πr^2 for collisions and there are n atoms per unit volume, the mean time, τ, between collisions is $1/(\pi r^2 n\sigma)$ and the line width is $2/\tau$: **17**

$$\text{Line width} = nfe^2/4\varepsilon_0 m\omega_0, \tag{4}$$

the oscillator strength f having been added in this expression. Various attempts to calculate the line width in this particular case have led to similar expressions, but with multiplying factors ranging between 1 and 8/3. Experimentally a factor of about 2 is indicated. The target radius must therefore be about 9 nm, much greater than the atomic radius. It is interesting to note that when sodium at low pressure is diluted with helium it requires 100 times the pressure of helium to reproduce the self-broadening resonance effect – the target radius is more like 0.9 nm. This is still rather larger than the distance of approach, say 0.3 nm, at which a direct collision could be said to occur, and one must infer that the interaction responsible for the van der Waals forces between non-polar molecules is responsible; the random fluctuating dipole moment of the helium atom, though changing on a time scale ten times as fast as the radiation frequency, can at these close distances perturb the sodium atom enough to cause a significant phase shift.

The model of identical oscillators just analysed is not really appropriate to the resonance line-broadening in ammonia at such low pressures that the lines in fig. 18.8 are well resolved. Most collisions will be between molecules in different rotational states, and the very fact that the lines are resolved implies that enough time elapses before a given molecule is dephased for the others to have gone through several cycles at least. Thus it may be assumed that the colliding molecules are different and have random relative phase, and that it is the change of phase suffered by each separately that matters, rather than by the sum as in the previous calculation. On the other hand the duration of the encounter is much less than a cycle, and rather strong mutual perturbations are needed to generate effective dephasing. The small differences in frequency between the colliding molecules are therefore unimportant when it comes to calculating the phase shift, and we may safely assume them identical during the encounter. Consider then the two harmonic oscillators whose states of vibration are initially represented by the complex amplitudes A and B. The normal modes have amplitudes $(A+B)/\sqrt{2}$ and $(A-B)/\sqrt{2}$, and during the encounter they change phase at different rates, so that the upper mode finishes up θ ahead of, and the lower mode θ behind, the phase which the unperturbed oscillators would have reached. With normal mode amplitudes $(A+B)\,e^{i\theta}/\sqrt{2}$ and $(A-B)\,e^{-i\theta}/\sqrt{2}$ at the end of the encounter the two oscillators are left with amplitudes $\frac{1}{2}[(A+B)\,e^{i\theta} \pm (A+B)\,e^{-i\theta}]$, i.e. $A\cos\theta + iB\sin\theta$ and $B\cos\theta + iA\sin\theta$. So far as the assembly of $A-$ molecules is concerned, the random phase of B means that the mean amplitude after the encounter is $A\cos\theta$, the term $iB\sin\theta$ vanishing in the average. Similarly B is reduced to $B\cos\theta$ by the collision. It appears that $\theta = 1$ is a fair criterion for an effective collision. Since $\theta = t\Delta\omega$, the encounter must last for a time $1/\Delta\omega$, which is the same result as was found appropriate for the dephasing of the total moment of identical molecules. The details of the model used seem unimportant.

We may now look at the same problem from the point of view of two double-well systems, treated quantally, rather than classical oscillators. The encounter is modelled by laying the two systems side by side and allowing their interactions to be switched on for a certain time. If the co-ordinates of the two systems are y_1 and y_2 the arrangement is now represented, as in fig. 2(a), by a single particle in a potential consisting of four square wells

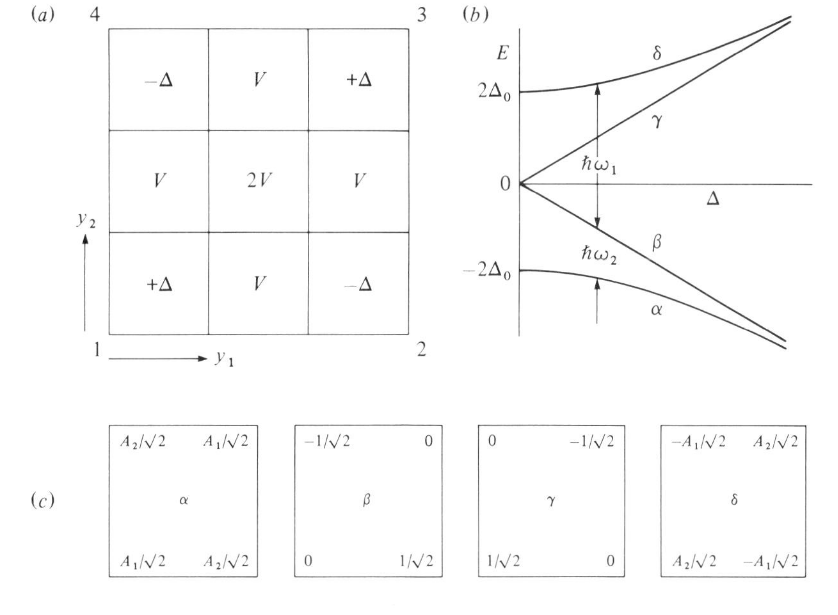

Fig. 2. (a) Potential energy for two double-wells coupled by dipole interaction, as measured by Δ; the wave-function is characterized by the amplitude of ψ in the wells at the four corners of the diagram.
(b) Perturbation of energy levels by Δ.
(c) Amplitudes of ψ in the four wells of (a) for the four stationary states of (b).

separated round the sides of the square by slightly penetrable barriers, the same as for the isolated double well, and with a central impenetrable region where the barrier is twice as high. The wells are taken as equal when the systems are isolated, but when they interact it is more favourable for opposite, rather than adjacent, wells to be occupied; each system may be supposed when polarized to produce a field disturbing the equality of the wells in the other. This is expressed by the potential $+\Delta$ in wells 1 and 3, and $-\Delta$ in 2 and 4. The energy levels are shown in fig. 2(b) and the corresponding amplitudes in the four wells in fig. 2(c). When $\Delta = 0$ the levels β and γ represent one system in its symmetrical ground state and the other in the antisymmetrical excited state; the degeneracy of those two states of the combined system is removed by the coupling, while at the same time the outer levels, which in the uncoupled condition represent both in the lower or both in the upper state, are further separated by the interaction. In the latter states the cosh and sinh functions in the barriers are displaced by the same y_0, and the ratio A_1/A_2 depends on y_0 exactly as in the isolated systems; the presence of a tunnelling barrier on two sides of each well, however, doubles the kinetic energy shift. Thus instead of (18.15), we have that

$$E_{\alpha,\delta} = \pm(4\Delta_0^2 + \Delta^2)^{\frac{1}{2}} \tag{5}$$

but A_1^2/A_2^2 is still given by (18.17). In the states β and γ the wave-function can be assumed to decay exponentially in going away from an occupied well; what minute amplitude reaches the next well is nullified by what arrives from the opposite corner. Since the wave-function is so nearly exponential there is no correction to the kinetic energy, and the levels are therefore at $\pm\Delta$, as in the diagram.

Let us suppose that at the moment of coupling $\Psi(y_1, y_2)$ is an arbitrary superposition of the four states, with amplitudes α, β, γ and δ. Then the amplitudes, B_1 to B_4, in the four wells, numbered as in the diagram, are as follows:

$$B_{1,3} = (\alpha A_1 \pm \gamma + \delta A_2)/\sqrt{2}, \qquad B_{2,4} = (\alpha A_2 \pm \beta - \delta A_1)/\sqrt{2}, \tag{6}$$

the upper signs applying to B_1 and B_2. After a time t each coefficient must be multiplied by its appropriate phase factor, $e^{-iE_\alpha t/\hbar}$ etc. The probability of finding the particle of the first system in its left-hand well is the probability that either of wells 1 and 4 is occupied, and if we write p_1 for the dipole moment of this system,

$$\begin{aligned}\langle p_1\rangle/p_0 &= B_2B_2^* + B_3B_3^* - B_1B_1^* - B_4B_4^* \\ &= A_2(\alpha\beta^* - \gamma\delta^*)\,e^{i\omega_2 t} - A_1(\beta\delta^* + \alpha\gamma^*)\,e^{i\omega_1 t} + \text{c.c.}\end{aligned} \tag{7}$$

in which $\hbar\omega_1 = E_\gamma - E_\alpha = E_\delta - E_\beta$ and $\hbar\omega_2 = E_\beta - E_\alpha = E_\delta - E_\gamma$. This is the general result which we specialize by assuming that at the moment of coupling Ψ can be written as a product function $[a_1\chi_1(y_1) + a_2\chi_2(y_1)]$ $[b_1\chi_1(y_2) + b_2\chi_2(y_2)]$, so that

$$B_{1,3} = \tfrac{1}{2}(a_1 \mp a_2)(b_1 \mp b_2), \qquad B_{2,4} = \tfrac{1}{2}(a_1 \pm a_2)(b_1 \mp b_2). \tag{8}$$

By comparing (6) and (8) the amplitudes α etc. are determined in terms of a_1, b_1 etc. After a certain amount of algebra and use of (18.12), (7) emerges in the form:

$$\langle p_1\rangle/p_0 = \mathrm{Re}\,[\{A_2^2(\xi_1+\mathrm{i}\eta_2\zeta_1)+A_1A_2(\xi_2\zeta_1+\mathrm{i}\eta_1)\}\,\mathrm{e}^{\mathrm{i}\omega_2 t}$$
$$+\{A_1^2(\xi_1-\mathrm{i}\eta_2\zeta_1)-A_1A_2(\xi_2\zeta_1-\mathrm{i}\eta_1)\}\,\mathrm{e}^{\mathrm{i}\omega_1 t}], \quad (9)$$

in which (ξ_1, η_1, ζ_1) and (ξ_2, η_2, ζ_2) are the positions of the representative points in fig. 18.1 for the two systems at the start of the interaction. In the case where the second system arrives with its dipole oscillation randomly phased, the average effect on $\langle p_1\rangle$ is obtained by putting $\bar{\xi}_2=\bar{\eta}_2=0$, and

$$\langle \bar{p}_1\rangle/p_0 = \mathrm{Re}\,[(A_2^2\xi_1+\mathrm{i}A_1A_2\eta_1)\,\mathrm{e}^{\mathrm{i}\omega_2 t}+(A_1^2\xi_1+\mathrm{i}A_1A_2\eta_1)\,\mathrm{e}^{\mathrm{i}\omega_1 t}]. \quad (10)$$

When the interaction is weak, A_1 and A_2 differ only slightly from $1/\sqrt{2}$ and the principal effect arises from the beating of the two terms in (10), whose frequencies are $(2\Delta_0 \pm \Delta)/\hbar$ to first order in Δ. Then

$$\langle \bar{p}_1\rangle/p_0 \approx \mathrm{Re}\,[\rho_1\,\mathrm{e}^{2\mathrm{i}t\Delta_0/\hbar}]\cos(t\Delta/\hbar). \quad (11)$$

This shows the same form of behaviour as do two classical oscillators; the amplitude of the oscillatory dipole is altered by a factor $\cos(\Delta\omega t)$, where in this case $\Delta\omega=\Delta/\hbar$. The results are in fact identical, since for two oscillators lying side by side $\Delta\omega = fe^2/8\pi\varepsilon_0 mr^3\omega_0$, and if f is given the value $4m\Delta_0 p_0^2/\hbar^2 e^2$ appropriate to the double-well system, $\Delta\omega = p_0^2/4\pi\varepsilon_0\hbar r^3$ which is $\Delta/\hbar$ for two double wells in the same configuration. The classical equivalent oscillator is therefore, as we have now come to expect, a satisfactory model for this case of weak interaction.

This result does not hold when the interaction is strong, which is hardly surprising since we have already seen that the equivalent oscillator does not describe strong perturbations of the two-level system. The extreme case of strong interaction occurs when the mutual polarization causes Δ to become much greater than Δ_0, so that $A_1 \sim 1$, $A_2 \sim 0$; then $\omega_1 \sim 2\Delta/\hbar$ and (10) reduces to the expression

$$\langle \bar{p}_1\rangle/p_0 = \xi_1\cos(2t\Delta/\hbar). \quad (12)$$

Since $\Delta \gg \Delta_0$ the perturbation needs to act for only a fraction of a cycle to be effective, and $\hbar/2\Delta$ may be taken as a measure of the interaction time required. This differs by only a factor of 2 from the criterion that applies when the interaction is weak.

It is the strong interaction result (12) that is most appropriate to the broadening of the microwave lines in ammonia. Let us ignore for the moment the rotation of the molecules and use the same approximations as before to estimate the critical distance of approach. Since $\Delta = p_0^2/4\pi\varepsilon_0 r^3$ and the encounter time is r/v, a value for r of $(p_0^2/2\pi\varepsilon_0\hbar v)^{\frac{1}{2}}$ is required to make $2t\Delta/\hbar$ equal to unity. Hence the target area is $p_0^2/2\varepsilon_0\hbar v$ and the collision rate when there are n molecules per m^3 is $p_0^2 n/2\varepsilon_0\hbar$, i.e. $1.3\times 10^{-14}\,n$. This may be compared with the experimental observations on the lines in fig. 18.8(c) for which $K=J$ and which approximate most closely to the model of non-rotating molecules. The line width is proportional to pressure and at a pressure of 0.5 torr is about 1.5×10^7 Hz, equivalent to a collision rate of $4.7\times10^7\ \mathrm{s}^{-1}$. At this pressure n is about $1.7\times10^{24}\ \mathrm{m}^{-3}$, so that the experimentally determined relationship is that the collision rate is $2.8\times10^{-14}\,n$. As with the estimate for the optical line of sodium the theoretical estimate is too low, but only by a factor of about 2, which is certainly within the limits of error of the estimate.

The discrepancy is really rather greater than has been suggested, perhaps by another factor of 2, because the rotation of the molecules has been ignored. When the quantum numbers are J and K, the total angular momentum is $\hbar[J(j+1)]^{\frac{1}{2}}$, of which $\hbar K$ is the component along the axis, as shown in fig. 1. The rotational speed is much higher than the inversion frequency, so that only a fraction $K/(J^2+J)^{\frac{1}{2}}$ of the dipole moment along the axis plays a significant role in the coupling between two molecules. The mean value of the steady dipole strength of the molecules colliding with a given one is therefore less by perhaps a factor of 2, and the line broadening to be expected is similarly reduced. The same argument implies that molecules having K equal, or nearly equal, to J should be especially sensitive to collisions, and those with small K less so; in fact the simple arguments given here suggest that the line width should be proportional to $K/(J^2+J)^{\frac{1}{2}}$. The measurements of Bleaney and Penrose are plotted in fig. 3 to show the strong correlation of line-width with $K/(J^2+J)^{\frac{1}{2}}$, but it

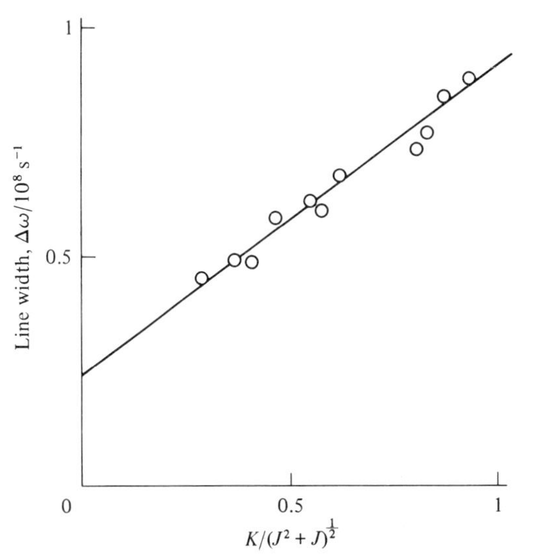

Fig. 3. Linear variation of line-width in the microwave spectrum of ammonia (fig. 18.8) when plotted against $K/(J^2+J)^{\frac{1}{2}}$.

is only fair to remark that they expected, on the basis of an equally rough-and-ready theory,[4] that the width should vary as $[K/(J^2+J)^{\frac{1}{2}}]^{\frac{2}{3}}$, and found in their results quite as satisfying a confirmation. It is not necessarily a refutation of the interpretation given here that the straight line does not pass through the origin. This may only mean that there are residual interactions between two ammonia molecules even when one of them has no steady dipole component; and the fact that some non-polar gas molecules can produce considerable broadening of the ammonia lines testifies to this. Let us therefore recognize that the line broadening processes can be fairly well understood in terms of simple models, even simple classical models, but that to take the argument further is no trivial matter. We therefore leave the question of what causes line broadening and proceed to the practical problem of how to get round it.

Doppler broadening; saturation spectroscopy

There is no need to say more about the origin of Doppler broadening. What is of interest here is the ingenious technique of saturation spectroscopy[5] by which its effects may be largely eliminated and the line-width reduced to something like the natural width, with consequent improved accuracy in wavelength measurement. It may be taken for granted that in such a measurement, aiming at the highest precision, the gas pressure is kept low enough for collision broadening to be unimportant, and that stray electric and magnetic fields are eliminated as far as possible, so that the natural width and the Doppler effect are all that are left to be contended with. There is not much to be done about the natural width, but the Doppler broadening is well worth removing. For example, the wavelengths of atomic hydrogen lines, such as $2P_{\frac{1}{2}}-3D_{\frac{3}{2}}$ at a wavelength of about 656 nm, are needed to very high precision for the determination of fundamental constants. This particular line has a theoretical natural life-time of 16 ns, corresponding to a Q-value of 4.5×10^7, so that a line-of-sight velocity for the atom of c/Q, or 7 ms^{-1}, would shift the line by as much as its width. With typical atomic velocities of 2 kms^{-1} the advantage to be gained is obvious.

The technique depends on the availability of tunable lasers giving enough power to cause appreciable saturation of the line by heating. As (18.56) shows, the value of $|\zeta_0|$, which determines the excess of unexcited over excited atoms, is reduced by irradiation, especially if the radiation is tuned to the resonance frequency of the transition. When the laser frequency is close to the transition frequency those atoms having the correct line-of-sight velocity will resonate exactly to the Doppler-shifted light and will be strongly excited; with strong irradiation, indeed, ζ_0 may be reduced nearly to zero, and the excited and unexcited populations will be nearly equal. Atoms moving with other line-of-sight velocities will be less affected. Another, and weaker, light beam at the same frequency, injected in the reverse direction, will for its part be absorbed most strongly by those atoms which are moving with the opposite velocity to those most affected by the first beam. If, then, the laser is detuned from the resonance frequency the atoms which are strongly affected by the first beam are not those that will absorb the second, and the absorption coefficient for the second will be hardly affected. On the other hand, when the laser is perfectly tuned it is the atoms at rest, or moving in the plane normal to the light, that are saturated by the first beam and are exactly those atoms that would absorb the second beam if they were not saturated. It is the disappearance of absorption that signals the condition of exact tuning.

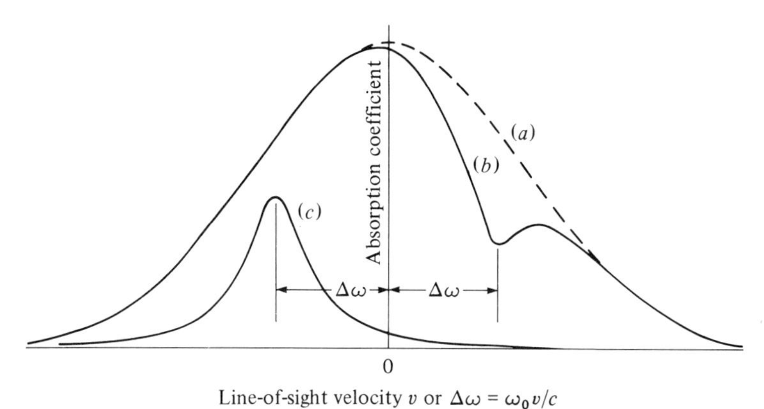

Fig. 4. Illustrating the principle of saturation spectroscopy; for description see text.

In fig. 4, curve (a) shows the line-of-sight velocity distribution, $f(v) \propto e^{-mv^2/2k_BT}$. When irradiated by light at a frequency $\Delta\omega$ above the resonance frequency, only those atoms moving away from the source with velocity close to $c\,\Delta\omega/\omega_0$ will absorb. If the laser power is weak, curve (a) represents the absorption coefficient of the gas as a function of frequency when the abscissae are scaled by ω_0/c to convert v into $\Delta\omega$. With more power, atoms moving in a band of velocities equivalent to something like the natural line-width suffer saturation and their absorption is reduced, as shown by curve (b) for $\Delta\omega > 0$. Now a weak beam from the same laser, but travelling in the opposite direction, is absorbed by atoms whose velocity is $c\,\Delta\omega/\omega_0$, also in the opposite direction. Since atoms which are not exactly in resonance, but lie within the natural line width, can absorb according to a Lorentzian formula, the total absorption for the weak beam is obtained by multiplying curve (b) by the Lorentzian (c), centred on $-\Delta\omega$, and determining the area under the product curve. Its variation with $\Delta\omega$ follows the general shape of the Doppler-broadened curve except when $\Delta\omega$ is so small that the dip in (b) and the hump of (c) overlap. It is this narrow dip that now constitutes the measured line-shape. The Doppler width, being so much greater, is virtually irrelevant except as a small correction factor in the analysis of the results; to develop the theory quantitatively we shall assume curve (a) to be flat except for the dip due to irradiation.

For an optical experiment conducted at room temperature ζ_0 may be taken to be -1, i.e. only spontaneous emission matters, and the representative point for a typical irradiated atom has co-ordinates ρ and ζ as given by (18.55) and (18.56); it describes a circular orbit of radius ρ_0 at a level

$\bar{\zeta}$, where

$$\bar{\zeta} = -1/[1 + I/(1 + \delta_+^2 \tau_p^2)]. \tag{13}$$

Here I is written for $p_0^2 \mathscr{E}_0^2 \tau_e \tau_p / \hbar^2$ and δ_+ for $\Delta\omega + \omega_0 v/c$, the mistuning of the Doppler-shifted radiation as seen by the moving atom. The spinning co-ordinate ρ is phase-linked to the radiation but this is irrelevant to the second beam which, travelling in the reverse direction, catches atoms in all phases with equal probability. Any individual atom experiences the second beam at a frequency mistuned by δ_-, i.e. $\Delta\omega - \omega_0 \sigma/c$, and the two beams combine to give an oscillatory field slightly modulated at the beat frequency $2\omega_0\sigma/c$. If the beams are both strong the modulation will cause periodic variations of $\bar{\rho}$ and lead to non-linear mixing, but with a weak second beam it is correct, to first order, to treat $\bar{\zeta}$ as constant at the value (13) determined by the first beam, and to use (18.55) to describe the response to the second beam:

$$\rho_0 = -\mathrm{i}C\mathscr{E}_2/[1 + I/(1 + \delta_+^2 \tau_p^2)][1 + \mathrm{i}\delta_- \tau_p], \tag{14}$$

where $C = p_0 \tau_0 / \hbar$ and $\mathscr{E}_2$ is the electric field strength in the second beam. The loss per atom is determined by Im $[\rho_0/\mathscr{E}_2]$ and we must find the total contribution by atoms moving at different speeds. The maximum value of Im $[\rho_0/\mathscr{E}_2]$ is $-C$, and occurs when the laser is perfectly tuned, the intensity weak, and $v = 0$; then $\delta_+ = \delta_-$, and $I \ll 1$. If we form W, defined by the expression,

$$W(\Delta\omega) = -\frac{1}{C} \int_{-\infty}^{\infty} \mathrm{Im}\, [\rho_0/\mathscr{E}_2)\, \mathrm{d}v,$$

W is a measure of the attenuation of the second beam at a mistuning of $\Delta\omega$. It is expressed as an equivalent velocity band-width – the attenuation is the same as if all atoms in the range $\pm W$ absorbed at the maximum value and the rest not at all. From (14),

$$W(\Delta\omega) = \int_{\infty}^{\infty} \mathrm{d}v (1 + \delta_+^2 \tau_p^2)/(1 + \delta_+^2 \tau_p^2/\alpha^2)(1 + \delta_-^2 \tau_p^2), \text{ where } \alpha^2 = 1 + I,$$

$$= \frac{\pi c}{\omega_0 \tau_p} \left\{ 1 - \frac{\alpha^2 - 1}{\pi \alpha^2} \int_{-\infty}^{\infty} \mathrm{d}x/[1 + (x_0 + x)^2/\alpha^2][1 + (x_0 - x)^2] \right\},$$

in which $x = \omega_0 \tau_p v/c$ and $x_0 = \tau_p \Delta\omega$. The integral is easily evaluated by taking a contour along the real axis and back round the semicircle at infinity. Then

$$W(\Delta\omega) = \frac{\pi c}{\omega_0 \tau_p} \{1 - (1 - 1/\alpha)/(1 + \Delta\omega^2 \tau_1^2)\}, \tag{15}$$

in which $\tau_1 = 2\tau_p/(1 + \alpha)$. At low intensities of the first beam, $(1 - 1/\alpha) \ll 1$, the dip in attenuation is small and the width is the natural line-width determined by τ_p. As I is increased the dip becomes stronger but at some cost in terms of line-width. The physical reason is obvious – the stronger beam affects $\bar{\zeta}$ over a wider range of frequency or velocity, and a really strong beam can saturate the transition over as wide a range as one may choose.

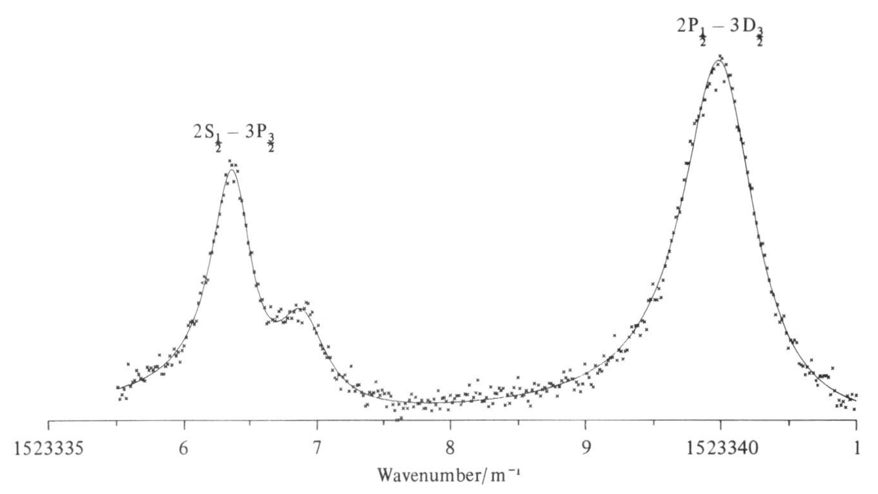

Fig. 5. Fine structure in the hydrogen spectrum resolved by saturation spectroscopy (Petley and Morris[6]). The full curve is theoretical, with the adjustable parameters chosen to give the best fit; from these parameters the wavenumber of each component can be determined with a precision of 0.01 ppm. The separation of the principal peaks (*Lamb shift*) is about 2.4 ppm, less than one-tenth of the Doppler width of each line.

The power of the technique is well illustrated in fig. 5, showing resolution of three fine-structure components of the hydrogen spectrum that would be lost in the Doppler profile.[6] The two principal components, separated by the Lamb shift, differ in frequency by 2 parts in 10^6; the Doppler profile for atomic hydrogen at room temperature is something like ten times wider than this.

Spin echoes and related effects

Our final example concerns the elimination of line broadening due to inhomogeneities, and we shall discuss two very different cases: nuclear spin resonance where the natural width of the lines is extremely small so that very small variations of magnetic field cause appreciable broadening; and double-well systems in glasses where the inhomogeneity is so great that no lines are normally detectable. The origin and use of echoes in such circumstances is conveniently discussed in terms of proton spin resonance, making use of the classical representation of the spin orientation as a point on a sphere. Since this carries over directly to any two-level system, the double well in particular, the second type of problem can be understood in principle as a straightforward extension of the first.

We shall be concerned with thermally equilibrated assemblies, every member of which is represented by a point on the sphere, with the centroid at some point ζ_0 on the vertical axis. In the absence of dissipation, and so long as the systems are identical, they respond to external forces in such a way that their centroid remains on a sphere of radius $|\zeta_0|$. It is convenient to replace the unit sphere of figs 18.1 and 18.4 by this smaller sphere, so that the lower pole represents the point to which the centroid tends as a result of dissipative effects; as before, the time-constant T_1 for relaxation in the ζ-direction is not normally the same as T_2 for relaxation in the ξ-η plane. The centroid provides all the information needed to interpret observations of the magnetic moment of a strictly homogeneous assembly of spins. If, however, the magnetic field varies with position in the sample, the spins precess at different rates and a cloud of points is now needed, each point representing the centroid of the spins in a region small enough to be treated as homogeneous.

I.224 Let us consider the effect of inhomogeneity on a typical nuclear induction experiment.[7] At the start all the points lie at the lower pole, but irradiation at the resonance frequency causes them to rotate about some horizontal axis, which we take as the ξ-axis, and in the ideal experiment irradiation is stopped when they have moved through 90° and reached the equator. In the frame of reference spinning at the resonant frequency they would all remain at the same point if the field were uniform; in the laboratory frame they would spin about a vertical axis to give a nuclear induction signal, constant in amplitude until dissipative effects supervened. Inhomgeneity of the steady field, however, causes different points to spin at different speeds; in a reference frame spinning at the mean speed the points fan out along the equator, forwards and backwards, causing their centroid to move in towards the axis. Without any dissipative process, therefore, the signal diminishes until, when the points are spread evenly around the equator, it disappears entirely. It can, however, be recovered by irradiation for a second period, twice as long as the first, so as to cause each point to turn through 180° about the ξ-axis. After an interval equal to that between the two irradiations the fan has closed up again and the induction signal is restored, only to fade away again as the fanning process continues. The reconstructed signal is the spin echo, and the diagram of fig. 6 illustrates

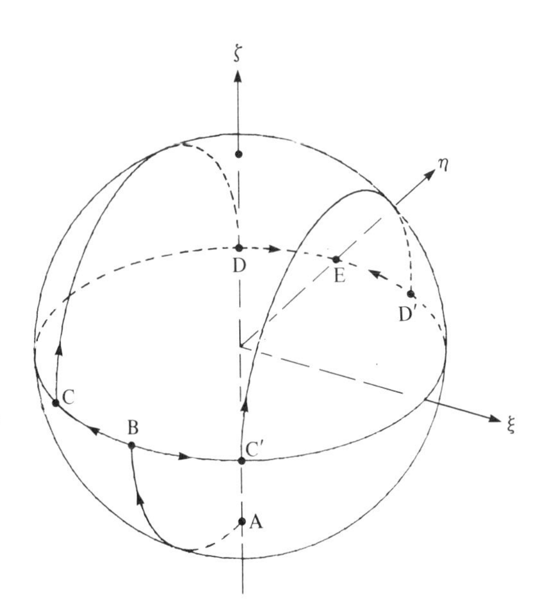

Fig. 6. Illustrating spin echo. A 90°-pulse takes the centroid of the spin assembly from A to B, where the representative points fan out round the equatorial plane; C and C′ show where two such points are at the instant a 180°-pulse is applied to take them to D and D′. After a time all points momentarily coincide at E. Note that the axes in this diagram are turned through 90° relative to fig. 7 and to fig. 18.1

the successive stages; this may also be expressed in Cartesian co-ordinates by carrying through the various rotations for a typical point which drifts through an angle θ during the fanning process:

A Initial state: $\zeta_0(0, 0, -1)$

B After 90° rotation about ξ-axis: $\zeta_0(0, -1, 0)$

C After fanning through θ about ζ-axis: $\zeta_0(-\sin\theta, -\cos\theta, 0)$

D After 180° rotation about ξ-axis: $\zeta_0(-\sin\theta, \cos\theta, 0)$

E After fanning through θ about ζ-axis: $\zeta_0(-\sin\theta\cos\theta + \cos\theta\sin\theta, \cos^2\theta + \sin^2\theta, 0)$, i.e. $\zeta_0(0, 1, 0)$.

So long as there is no random spreading of the points by dissipative effects, every point finishes up at E, as illustrated by the two trajectories in the diagram. At its momentary peak the echo is as strong as if there were no field inhomogeneity. Decay of the echo strength when longer intervals are allowed to elapse between the first 90° pulse and the second 180° pulse may be attributed to real dissipation, and thus the true time-constant may be separated from the artefacts of inhomogeneity. In an induction experiment it is the transverse component that is measured, and therefore T_2 controls the rate of echo decay.

Two matters are worth comment to supplement this brief account. First, it is technically easier to use identical pulses rather than have the second twice as long as the first. Let us follows the fortunes of a typical point when each pulse turns it through ε about the ξ-axis:

A Initial state: $\zeta_0(0, 0, -1)$

B After the first pulse: $\zeta_0(0, -\sin\varepsilon, -\cos\varepsilon)$

C After fanning through θ: $\zeta_0(-\sin\varepsilon\sin\theta, -\sin\varepsilon\cos\theta, -\cos\varepsilon)$

D After the second pulse:

$$\zeta_0(-\sin\varepsilon\sin\theta, -\sin\varepsilon\cos\theta\cos\varepsilon - \cos\varepsilon\sin\varepsilon, -\cos^2\varepsilon + \sin^2\varepsilon\cos\theta)$$

E After fanning through θ:

$$\zeta_0(-\sin\varepsilon\sin\theta\cos\theta - \cos\varepsilon\sin\varepsilon\sin\theta(1+\cos\theta),$$
$$-\cos\varepsilon\sin\varepsilon(\cos^2\theta+\cos\theta)+\sin\varepsilon\sin^2\theta, -\cos^2\varepsilon+\sin^2\varepsilon\cos\theta).$$

There is no question now of all points finishing at the same place, and inevitably some signal is lost. But if ε is well chosen the loss is not too great. Let us consider a common situation where the inhomogeneity causes signal loss in a time much shorter than T_1 or T_2, so that after the first pulse the points spread all round the sphere before being brought together again by the second. If all values of θ are equally represented, the final position of the centroid is obtained by averaging over θ putting $\cos\theta = \sin\theta = 0$, $\cos^2\theta = \sin^2\theta = \frac{1}{2}$, it is therefore to be found at $\zeta_0(0, -\frac{1}{4}\sin 2\varepsilon + \frac{1}{2}\sin\varepsilon, -\cos^2\varepsilon)$. The induction signal is determined by the transverse component, i.e. $\zeta_0(\frac{1}{2}\sin\varepsilon - \frac{1}{4}\sin 2\varepsilon)$, and is greatest when $\varepsilon = 120°$; its value is then $0.65\,\zeta_0$, rather than ζ_0 as with thc 90° + 180° pulses. The decay rate of the echo is still governed by T_2.

The second point concerns the degree to which the two pulses must be phase-coherent. The spin responds to a horizontal magnetic field rotating at the resonant frequency in the same sense as its precession, and in the rotating frame of reference we consider only this component of the oscillatory applied field, representing it as pointing in a direction in the horizontal plane that is determined by the phase of the oscillation. It is about this axis, which was taken as the ξ-axis in the above account, that the point rotates as a result of the irradiation. The assumption that the same axis applies to the effect of both pulses is tantamount to supposing that the oscillator remains switched on, at exactly the right frequency, between the two pulses, which are applied by gating the oscillator output at the appropriate moments. In fact, this rather delicate technique is not normally required, and it is enough to switch the oscillator on for each pulse separately, not worrying about the phase relationship. The two rotations through ε are now about different horizontal axes, and the effect of this is most readily seen graphically. In fig. 7 the original process is shown in perspective and

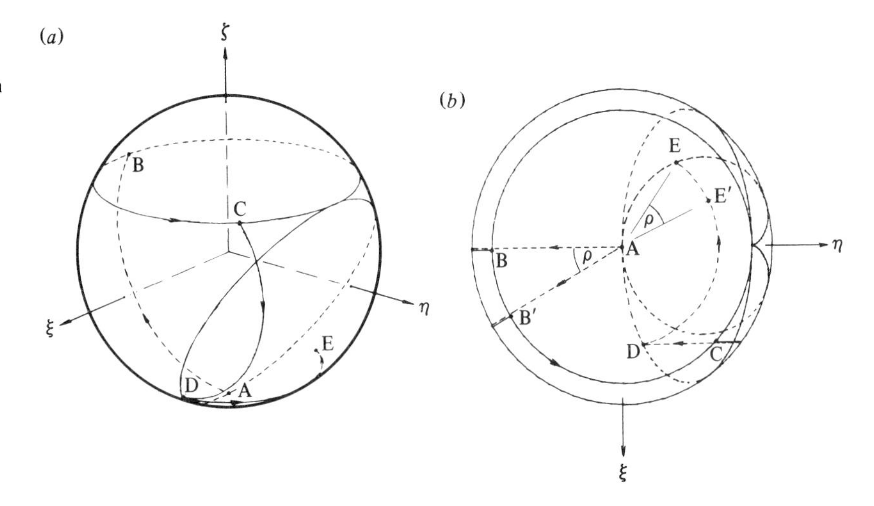

Fig. 7. Trajectory of a representative point in a spin echo experiment with two 120° pulses. (*a*) perspective view, (*b*) plan view. The trajectory ABCDE is typical for coherent pulses, while AB′CDE′ is typical for pulses with a relative phase shift of ϕ.

in plan, with $\varepsilon = 120°$, and as in fig. 6 the path followed by one system (with $\theta = 135°$) is shown in detail. When enough time is allowed for the points to make many revolutions and to spread evenly round the horizontal circle through B the second pulse, which tips the circle through 120°, gives rise to an induction signal since there is now a non-vanishing mean transverse moment, but this disappears as all the points fan out at different speeds, only to come together at the echo which arrives as the distribution collapses momentarily onto the curve which in plan resembles a cardioid. It is this that takes the place of the single point E in fig. 6.

Now let us suppose that the first rotation was about a different axis, so that points originally at A were shifted to B′. Then the system whose fortunes we followed from B to C, to D and ultimately to E is now most nearly matched in behaviour by another system whose fanning angle is $\theta - \phi$. This goes from A to B′ and then to C, whence it follows the same pattern as the original in going to D. However, in the last stage it does not fan round to E, but only to E′, since its fanning angle is smaller by ϕ. And the same may be said of every point, so that at the moment of echo all points lie on an identical cardioid, but rotated through ϕ. So long as there was time for the points to cover the circle completely, the cardioid will be covered in exactly the same way for all ϕ, so that only the phase, not the amplitude, of the echo is affected by lack of coherence between the pulses.

The echo produced as just described, by the use of two pulses, is known as the *spontaneous echo* to distinguish it from the *stimulated echo* for which three unequal pulses are required. The first is of such strength and duration as to turn the representation through 180°, so that the centroid which started at the south pole, $-\zeta_0$, is now at the north pole, $+\zeta_0$. There is no induction signal since no mean rotating dipole is created. The centroid relaxes back towards $-\zeta_0$ with the longitudinal relaxation time T_1, and at some stage before this process is complete two more pulses are applied as in the normal spontaneous echo experiment. As a result an echo is produced as long after the third pulse as the second was before it, and the amplitude is determined by the value, ζ_0', to which the centroid had relaxed by the time of the second pulse, and by the delay between the second and third. By keeping the latter delay constant and varying the time between the first and second pulses the relaxation of ζ_0' can be followed; the amplitude falls to zero as ζ_0' goes through the centre of the sphere and rises again, with phase reversal, to a level determined by the spontaneous echo strength. The combination of the two echo techniques enables T_1 and T_2 to be separately measured, free from the masking effect of field inhomogeneity.

The stimulated echo principle has been applied to study the relaxation time $\tau_e(\equiv T_1)$ of two-level systems in glasses. Although there is considerable uncertainty about the configuration responsible, there is little doubt that in many amorphous materials, especially silicate glasses, there is a rather high concentration of structures with all the essential characteristics of double-well systems. Most likely there is some commonly occurring atomic arrangement which provides single protons with the choice of two similar potential wells, close enough together to permit tunnelling. But these are not sparsely distributed and highly symmetrical like the proton systems in polyethylene; all inequalities of wells are to be found so that, even if they had the same Δ_0, the spread of Δ and hence of E would smear out any resonance line that might otherwise be detected. The ensemble of systems thus bears a certain resemblance to the spin system just discussed, and can be represented in the same way, but in place of a small degree of field inhomogeneity and consequent small spread of E there is here a very wide spread. The echo technique, however, allows systems whose values of E lie in a narrow band to be selected and their relaxation time measured. Since this is related to the tunnelling probability and the coupling to lattice vibrations the information that is yielded in principle is the same as can be obtained from dielectric loss studies in polyethylene and similar materials where the confusing effects of inhomogeneity are almost absent. In glasses the hypothetical Debye relaxation peaks are smoothed out into an almost frequency-independent dielectric loss, from which little of interest can be gleaned.

A *photon-echo* experiment, analogous to the stimulated spin echo, has been performed, at very low temperatures, on samples of silica glass in a microwave cavity.[8] The electric field pulses used to shift the representative points round the sphere were applied as bursts at the cavity frequency 720 MHz, at which some at least of the double-well systems resonate. Let us suppose that all the systems have the same Δ_0, with a wide spread of Δ, and (for convenience) that the oscillator has been tuned to $2\Delta_0/\hbar$. Then the systems with $\Delta = 0$ will ideally be inverted by the first pulse, their representative points being turned through 180° about a horizontal axis. Those, however, with non-zero Δ will be mistuned by an amount $\Delta_0/E - 1$, which we denote by δ as in (18.4), and their points will turn at a rate $\delta' = (\delta_0^2 + \delta^2)^{\frac{1}{2}}$ about an axis tilted $\tan^{-1}(\delta/\delta_0)(=\theta)$ from the horizontal,

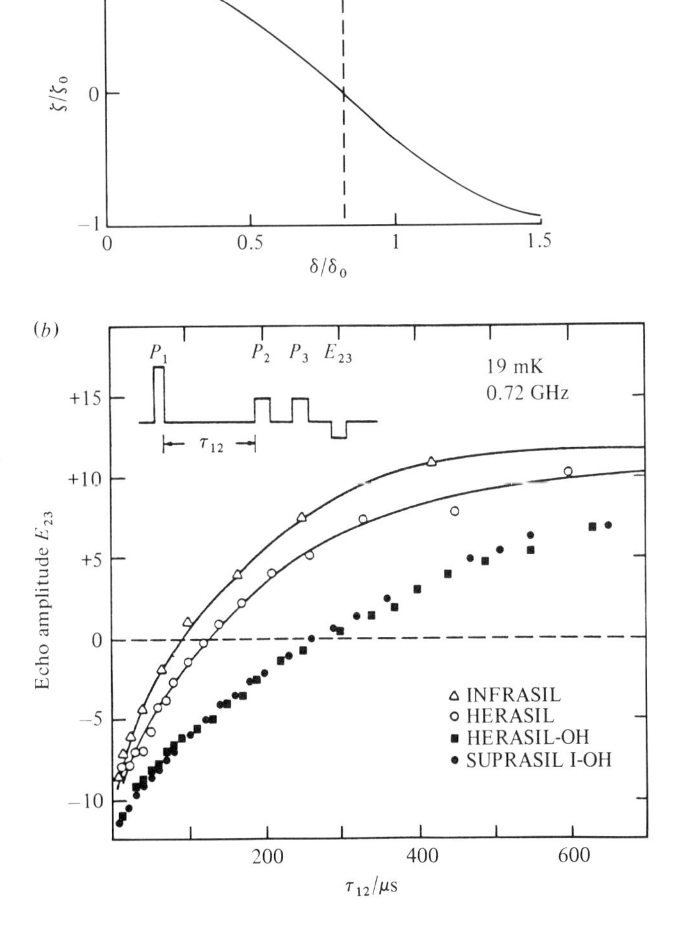

Fig. 8. (*a*) The centroid of representative points, initially at $-\zeta_0$, is raised to ζ by a 180° pulse; the curve shows how ζ is affected by mistuning between the frequency of the double-well and the pulse. The area under the curve is indicated by the broken-line rectangle of width 0.8. (*b*) Showing how T_1 may be determined in a photon echo experiment by the variation of echo amplitude with the time delay, τ_{12}, between the 180° pulse and the first 90° pulse[8]. The different points refer to different proprietary brands of silica glass.

as in fig. 18.1; here δ_0, is a measure of the strength of the oscillatory field. If the length of pulse is π/δ_0, a typical off-tune point is turned through $\pi(\delta_0^2+\delta^2)^{\frac{1}{2}}/\delta_0(=\beta)$, giving the ideal of π when $\delta=0$. Now a point initially at $-\zeta_0$, after turning through β about an axis tilted θ upwards from the ξ-axis, finishes at $\zeta_0(-\frac{1}{2}\sin 2\theta(1-\cos\beta), -\cos\theta\sin\beta, \cos^2\theta(1-\cos\beta)-1)$. The spread of E ensures that in practically no time the points have fanned out round the sphere, and we shall concern ouselves only with the ζ-component, which depends on δ/δ_0 as shown in fig. 8(a). The centroid of the assembly has been raised as much as if all those systems with $|\delta|<0.8\delta_0$ had been perfectly inverted and the rest left untouched. If the pulse is weak and appropriately long, δ_0 is small and there is sharp selectivity of systems for promotion. It is on these that the subsequent echo experiment is performed, the second and third pulses being close together since it is T_1 and not T_2 that is of interest. The experimental curves in fig. 8(b), taken at a temperature of 19 mK, show clearly the decay of ζ_0' from positive to negative, with the echo amplitude passing through zero. It is interesting that these experiments revealed clearly two different double-well systems, with different relaxation times (140 μs and 410 μs in the examples shown here) and requiring different pulse strengths to bring them to their maximum echo amplitudes. This is an excellent example of the use of ingenious technique to bring out significant features in what at first sight appears a featureless and uninteresting phenomenon.

20 The ammonia maser

The original maser of Gordon, Zeiger and Townes,[1] driven by a focussed stream of ammonia molecules in their antisymmetrical state, provides a conveniently explicit example on which to base a discussion of the principles underlying coherent excitation of a vibrator by stimulated emission. It was shown in chapter 18 how a quadrupole electrostatic lens served to separate symmetric from antisymmetric states, and we shall assume that separation is perfect; it is easy to extend the argument to include a proportion of molecules in the symmetric state. In addition we shall ignore any complications arising from the multitude of rotational states leading to the fine structure shown in fig. 18.8, and shall assume that only one line contributes, for example the strong 3,3 line at 23.9 GHz. Since the microwave cavity resonator, if it is to be excited by the molecules, must normally be very closely tuned to their natural frequency this assumption is realistic.

124

The simplest intuitive approach to the maser is by way of Einstein's treatment of radiation in terms of stimulated and spontaneous processes.† Excited molecules passing through the resonator, when it is already in an oscillatory condition, are stimulated by the field; if the resonator frequency is well matched to the molecular levels they may make a transition down to the ground state and on leaving the resonator have $2\Delta_0$ less energy than when they entered. The radiated energy is phase-coherent with the cavity vibration, whose amplitude is thereby increased; every molecule making a transition increases the quantum number of the vibration by one. The lifetime for spontaneous decay of the excited state of the molecule is so long in relation to the residence time in the resonator that spontaneous processes do not play a large role and may be ignored for the moment. The resonator walls are normally dissipative, and if the maser is used as a microwave source the extraction of power augments the natural dissipation. Since the rate of power transfer from a molecule to the resonator is proportional to the energy density in the resonator, and since the dissipation rate is also proportional to energy density, there is a critical flux of excited molecules below which the excitation is insufficient to overcome damping; for higher than critical flux, however, the excess power provided by the

79, 113

† If we refer to this approach as the Einstein argument, it is not meant to imply that he had any part in it except for the introduction of the radiation coefficients into the vocabulary of physics. Certainly he is not responsible for the fundamental flaws in the argument as applied to the maser.

molecules causes the level of oscillation to rise until the assumptions of the argument are invalidated, and the resonator settles down to a steady level. These processes have the appearance of being readily understood and formulated mathematically, as we shall show, but considerable refinement is called for if correct quantitative results are sought. A more classical approach is much less open to criticism, and we shall develop the response-function argument of chapter 18 for this purpose. The Einstein argument enables the threshold flux to be calculated and gives a fairly good picture of the steady state. It also, by an elementary application of the uncertainty principle, indicates how well tuned the cavity must be – if the residence time of a molecule is T, an error in frequency amounting to less than $1/T$ will hardly be noticed before the molecule has left. The response-function approach supplies everything in the Einstein argument, while showing much more explicitly by how much the critical flux increases with detuning, and hence it provides a figure for the frequency range within which maser action will occur for a given flux. Furthermore it shows that when the system is mistuned the frequency of steady oscillation is a compromise between that of the resonator and that of the molecule, with the latter strongly favoured. It does, however, fail to reveal the existence of a noise source which the fully quantal treatment shows to be present, not just as a concomitant of the dissipative process (Johnson noise) but equally as inseparable from the excitation. The effect of noise is principally to cause the phase of oscillation to drift randomly and thus to limit the purity of the output. In practice quantum noise is considerably less important than thermal noise for a maser operating in the microwave frequency range, but the way it emerges from the quantum mechanics is so intellectually satisfying that we shall develop the argument in detail.

Stimulated emission

From now on we shall adopt a simplified version of the resonator, assuming that along the line travelled by a molecule as it passes through, not only the phase but also the amplitude of electric field oscillation is constant rather than varying sinusoidally with position as befits a half-wavelength loop of a standing wave. If the energy in the resonator is E when this field reaches its maximum value $\mathscr{E}_0$, we write $\frac{1}{2}\varepsilon_0\mathscr{E}_0^2 V_{\text{eff}}$ for E and thus define the effective resonator volume, V_{eff}. The molecule moves in an electromagnetic field of energy density $u = \frac{1}{2}\varepsilon_0\mathscr{E}_0^2$, and if we take the Einstein-inspired arguments of chapters 16 and 18 at their face value we shall assume that an excited molecule will be stimulated to emit a quantum at a rate proportional to u, while an unexcited molecule will be stimulated to absorb at the same rate. In fact, when the resonator supports only one mode of interest this is not a valid point of view, for after an excited molecule has lost a quantum its wave-function is still coherent with the regular oscillation of the electromagnetic field, so that its subsequent behaviour is not independent of its past; and this is not the only failure of physical reasoning, as will become apparent shortly. Nevertheless, we ignore such niceties and proceed in naive hope. If at a time t after entering the resonator a fraction f of the molecules are still excited, we write the de-excitation rate $-\dot{f}$ as $B_{12}uf$; while as a result of re-excitation there is a contribution $B_{12}u(1-f)$ to $\dot{f}$. Hence

$$\dot{f} = B_{12}u(1-2f),$$

so that f relaxes exponentially towards the steady-state value $\frac{1}{2}$. In particular, if $f=1$ when the molecules enter, f varies during passage as $\frac{1}{2}(1+\mathrm{e}^{-2B_{12}ut})$, and the mean value of this expression at the moment of exit gives the fraction that have not communicated their energy to the resonator. Because of the inevitable spread of molecular velocities, the residence times vary widely, and it is not a very satisfactory approximation to insert the mean residence time T in the exponential, as we now do. However, the general form of the behaviour will not be seriously falsified.

Let us write Φ for the molecular flux, the number of excited molecules entering in unit time; then $\frac{1}{2}\Phi(1+\mathrm{e}^{-2B_{12}uT})$ emerge excited, the rest contributing $\hbar\omega_0$ to the resonator energy, or $\hbar\omega_0/V_{\mathrm{eff}}$ to u. Hence

$$\dot{u} = \tfrac{1}{2}\Phi(1-\mathrm{e}^{-2B_{12}uT})\hbar\omega_0/V_{\mathrm{eff}} - u/\tau_{\mathrm{e}}, \tag{1}$$

where τ_{e} is the decay time for energy in the freely oscillating resonator. By writing x for $2B_{12}uT$ and $F\Phi_{\mathrm{c}}$ for Φ, where $\Phi_{\mathrm{c}} = V_{\mathrm{eff}}/B_{12}T\hbar\omega_0\tau_{\mathrm{e}}$, (1) is cast in the form:

$$\tau_{\mathrm{e}}\dot{x} = F(1-\mathrm{e}^{-x}) - x, \tag{2}$$

from which the development, if any, of the energy of oscillation can be determined. When x is small, $\tau_{\mathrm{e}}\dot{x} \sim (F-1)x$, and it is clear that no growth can occur unless $F>1$; Φ_{c} as defined above is therefore the critical flux. In terms of Φ_{c}, $x = 2E/\hbar Q_{\mathrm{r}}\Phi_{\mathrm{c}}$, where $Q_{\mathrm{r}} = \omega_0\tau_{\mathrm{e}}$, the natural quality factor of the resonator.

In the steady state $\dot{x} = 0$ in (2). When $F-1 \ll 1$, x settles down to the value $x_{\mathrm{s}} = 2(F-1)$ and as F is increased x_{s} tends asymptotically to F. One obtains a clear picture of the scale of operation of a maser by calculating the mean quantum number $\bar{n}_{\mathrm{s}}$ of the resonator oscillation corresponding to a given value of x_{s}:

$$\bar{n}_{\mathrm{s}} = V_{\mathrm{eff}}u_{\mathrm{s}}/\hbar\omega_0 = V_{\mathrm{eff}}x_{\mathrm{s}}/2B_{12}T\hbar\omega_0 = \tfrac{1}{2}x_{\mathrm{s}}\tau_{\mathrm{e}}\Phi_{\mathrm{c}}. \tag{3}$$

In the original maser Φ_{c} was about 10^{13} molecules per second and τ_{e} about 10^{-7} s ($Q_{\mathrm{r}} \sim 10^4$), so that with $F=2$ and $x_{\mathrm{s}} = 1.6$, $\bar{n}_{\mathrm{s}} \sim 10^6$. In everything that follows it will be assumed that $\bar{n}_{\mathrm{s}}$ is very high, and this will permit occasional mathematical simplifications.

It is easy enough to integrate (2) numerically to show how x builds up from a small value when F is brought above unity – exponentially at first with a time-constant of the order of τ_{e}, then slower, and finally an exponential approach to x_{s} with a different time-constant, also of the order of τ_{e}. Unfortunately this argument overlooks the fact that the residence time T

is about 10^{-4} s, considerably greater than τ_e. It is therefore quite wrong to assume that u is sensibly constant during the passage of a molecule, and the rate is in fact determined more by T than by τ_e; this does not invalidate the expressions for Φ_c and x_s, which depend on the nature of the steady state. There is so much wrong with the basic arguments leading to (2) that it is not worth attempting to derive a better treatment of the non-stationary state. The next approach we shall develop is much better in principle and will be taken as far as studying fluctuations from the steady state where the conflict of two very different characteristic times has important consequences.

161

Dielectric response of a molecular beam

The second approach to the ammonia maser has already been adumbrated in chapter 11, but we are now in a position to make the argument quantitative. We have seen that the response functions of the unexcited and excited states are opposite in sign, so that if the former gives rise to dielectric loss the latter conversely gives rise to dielectric gain. The critical flux is such that the beam of molecules, considered as a dielectric rod in the resonator, is just able by its gain to neutralize the resonator losses. Linear response theory will take one thus far, but to reach a steady state of oscillation non-linearity must play a part. We are concerned to find the mean dipole moment of a typical molecule as it traverses the resonator, whose vibration frequency is not necessarily the same as the natural resonator frequency. Since a number of different frequencies and related quantities enter the theory, we shall list the most important before proceeding further:

I.347

- ω_0 natural frequency (real) of resonator, whose natural decrement is described by τ_e or $Q_r\,(=\omega_0\tau_e=\frac{1}{2}\omega_0\tau_a)$.
- ω_m natural frequency of the molecular transition $=2\Delta_0/\hbar$; the Q associated with spontaneous decay is enormous and will not concern us, but we shall introduce an effective Q related to the residence time T rather than to the lifetime: $Q_m=\frac{1}{6}\omega_0 T$.
- ω the actual frequency of the maser in its steady state.
- ω_e the mistuning of empty resonator and molecule, defined as $\frac{1}{2}(\omega_m-\omega_0)$.
- ω_r the resultant mistuning of maser and molecule, defined as $\frac{1}{2}(\omega_m-\omega)$.
- Ω_n a measure of the normal mode frequency shift when a molecule is coupled to the resonator in its nth excited state; according to (18.36), $\Omega_n^2=\omega_e^2+\gamma^2(n+1)$ where γ is the coupling constant.

To relate γ to the resonator characteristics, note that C in (18.31) gives Cx as the interaction energy between an oscillator with displacement x and the double-well system with the particle in one well. Hence Cx is equivalent to $\mathscr{E}_0 p_0$, while by considering energy we see that $m_0\omega_0^2 x^2$ is equivalent to $\varepsilon_0 V_{\text{eff}}\mathscr{E}_0^2$. Hence C^2 is to be interpreted as $m_0\omega_0^2 p_0^2/\varepsilon_0 V_{\text{eff}}$, and γ^2, which is $C^2/2\hbar\omega_0 m_0$, is $\omega_0 p_0^2/2\hbar\varepsilon_0 V_{\text{eff}}$.

We now introduce three dimensionless quantities:

$\Gamma = 2\omega_r T$,

$\Gamma_0 = p_0 \mathscr{E}_0 T/\hbar = 2\gamma n^{\frac{1}{2}} T$,

and $\Gamma' = (\Gamma^2 + \Gamma_0^2)^{\frac{1}{2}}$. For future reference it may be noted that $\Gamma' = 2\Omega_n T$ if ω_e is replaced by ω_r and the difference between n and $n+1$ is ignored.

The analysis accompanying fig. 18.1 contains what is needed to find the mean dipole moment of a molecule that enters in the excited state and experiences a very large number of cycles of resonator oscillation at frequency ω during residence. The point representing the momentary amplitude, and phase relative to the resonator oscillation, starts at the upper pole and moves for time T in a circular orbit on the sphere at angular velocity $(4\omega_r^2 + p_0^2\mathscr{E}_0^2/\hbar^2)^{\frac{1}{2}}$, i.e. Γ'/T, so that it traces out an arc of length Γ'. The inclination of the orbit, θ, is $\tan^{-1}(\Gamma/\Gamma_0)$ and the orbit radius is $\cos\theta$, i.e. Γ_0/Γ'. The centroid of the arc is easily found, and its projection on to the horizontal plane gives both the real ($\bar{\xi}$) and imaginary ($\bar{\eta}$) parts of the mean dipole moment in terms of p_0:

$$\bar{\xi} = -\Gamma\Gamma_0(\Gamma' - \sin\Gamma')/\Gamma'^3, \qquad \bar{\eta} = -\Gamma_0(1-\cos\Gamma')/\Gamma'^2. \tag{4}$$

There are ΦT molecules present at any instant, so that the total dipole moment P_{tot} is $\Phi T p_0(\bar{\xi} + \mathrm{i}\bar{\eta})$. This is not spread throughout the resonator in proportion to the local electric field, as the polarization of a uniform linear dielectric would be, but an energetic argument is readily devised to show that the effective volume susceptibility $\kappa_e\,(=\kappa_e' + \mathrm{i}\kappa_e'')$ is $P_{\text{tot}}/V_{\text{eff}}\varepsilon_0\mathscr{E}_0$. Hence, from (4),

$$\kappa_e = -(\Phi T^2 p_0^2/\varepsilon_0\hbar V_{\text{eff}}\Gamma'^2)[\Gamma(\Gamma' - \sin\Gamma')/\Gamma' + \mathrm{i}(1-\cos\Gamma')]. \tag{5}$$

The sign of κ_e' is opposite to that of Γ, negative when $\omega < \omega_m$. The lower state of the molecule has positive polarization at low frequencies but the upper state has negative. Similarly the sign of κ_e'' describes dielectric gain rather than the loss shown by the lower state.

Since $\mathscr{E}_0$ is involved in Γ', κ_e is obviously not independent of $\mathscr{E}_0$, except in weak fields; then Γ' can be replaced by Γ without significant error or, when the system is perfectly tuned, $1-\cos\Gamma'$ can be replaced by $\frac{1}{2}\Gamma'^2$. In the latter case $\kappa_e' = 0$ and $\kappa_e'' = -\Phi T^2 p_0^2/2\varepsilon_0\hbar V_{\text{eff}}$. The resonator behaves as if filled with a dielectric of relative permittivity $\varepsilon = 1 + \kappa_e$, and its resonant frequency is shifted from ω_0 to $\omega_0/\varepsilon^{\frac{1}{2}}$, i.e. $\omega \sim \omega_0(1 - \frac{1}{2}\kappa_e)$ when $\kappa_e \ll 1$. The imaginary part of ω is $-\frac{1}{2}\kappa_e''\omega_0$ and consequently the decay time for energy is $1/\kappa_e''\omega_0$. It follows that if the natural decrement of the resonator is described by a certain τ_e, the critical flux needed to overcome this is such that $\kappa_e'' = -1/\omega_0\tau_e = -1/Q_r$. Hence

$$\Phi_c = 2\varepsilon_0\hbar V_{\text{eff}}/Q_r p_0^2 T^2 = 1/\tau_e\gamma^2 T^2, \tag{6}$$

and (5) may be written:

$$\kappa_e = -(2F/Q_r\Gamma'^2)[\Gamma(\Gamma' - \sin\Gamma')/\Gamma' + \mathrm{i}(1-\cos\Gamma')], \tag{7}$$

with $F = \Phi/\Phi_c$, as in (2).

Before proceeding to discuss the consequences of this result, let us note that comparison of the two expressions for Φ_c, (6) and that following (1), shows what meaning should be ascribed to the stimulated emission coefficient B_{12}:

$$B_{12} = p_0^2 T / 2\varepsilon_0 \hbar^2.$$

The appearance of the residence time T in this expression casts grave doubt on the argument, since B_{12} was introduced to define the de-excitation rate of a molecule in a field of given energy density, with no thought of how long it was to remain in that field. What we have unjustly called the Einstein argument involved the assumption that the probability of de-excitation is initially proportional to the elapsed time after exposure to resonator field. This is only tenable when a large number of incoherent modes are simultaneously acting on the molecule. If there is only one mode the probability is initially proportional to t^2, and the intrusion of another time-like quantity into the expression can be seen to be required by dimensional considerations. Unfortunately, however appealing the Einstein argument may be to the physical intuition, it is deeply flawed as a quantitative procedure.

Returning to the dielectric approach, when the amplitude of oscillation is low enough for Γ' to be replaced by Γ, κ_e has the form shown in fig. 1.

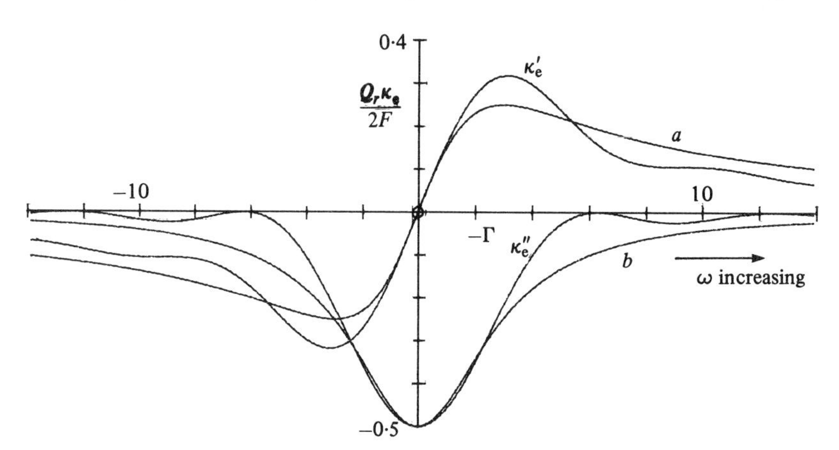

Fig. 1. Real (κ_e') and imaginary (κ_e'') parts of the susceptibility of a beam of excited ammonia molecules, compared with the corresponding curves (a and b) for a Lorentzian resonance adjusted to fit at $\Gamma = 0$.

The ripples on the wings, which result from the assumption that T is the same for all molecules, would be smoothed out in any real device. Even without smoothing, the general form of the curve is fairly close to the Lorentzian response of a simple resonant system, and agreement would probably be still better after smoothing. The Lorentzian curves shown for comparison are the real and imaginary parts of $-(\mathrm{i}F/Q_r)/(1+\frac{1}{3}\mathrm{i}\Gamma)$, which is chosen to match the slope of κ_e' and the magnitude of κ_e'' at $\Gamma = 0$. The width of this curve, $\Delta\Gamma$, is 6, so that the frequency width is $6/T$, corresponding to a Q-value of $\frac{1}{6}\omega_0 T$, which is what we have chosen to define Q_m, the quality factor of the molecular transition; Q_m is typically of the order of

10^6. It is the residence time, not spontaneous radiation, that determines the effective width of the quasi-resonant response. It will be observed that the unsmoothed form of κ_e'' goes to zero when Γ (or in general Γ') $= 2\pi$; this carries the implication that the oscillation cannot build up to the point where $\Gamma' = 2\pi$, and all points outside $\pm 2\pi$ are irrelevant. This would not be true for the smoothed curve.

As the amplitude builds up, non-linearities bring the system to a steady state in which κ_e'' is just sufficient to overcome dissipation in the resonator. The argument developed for linear response is still valid since κ_e'' has been defined, even in the non-linear situation, in such a way that $-\frac{1}{2}\kappa_e''\varepsilon_0\mathscr{E}_0^2 V_{\text{eff}}$ is $-\frac{1}{2}\varepsilon_0\mathscr{E}_0 \operatorname{Im}[P_{\text{tot}}]$, the power supplied by the molecular beam. Hence in the steady state $\kappa_e'' = -1/Q_r$ and, from (7),

$$2F(1-\cos\Gamma') = \Gamma'^2, \text{ or } F = (\tfrac{1}{2}\Gamma'/\sin\tfrac{1}{2}\Gamma')^2. \tag{8}$$

The value of Γ' is thus uniquely determined by F. It follows then from (7) that κ_e' in the steady state is uniquely determined by F and the mistuning; thus by carrying this argument through we are able to discover ω, the frequency of oscillation in the steady state. For the effect of κ_e', the real part of the susceptibility, is to change ω_0 to $\omega_0(1-\frac{1}{2}\kappa_e')$, and since $\omega - \omega_0 = 2(\omega_e - \omega_r)$ we have that

$$\omega_e - \omega_r = \tfrac{1}{4}\omega_0\kappa_e' = (F\omega_0\omega_r T/Q_r\Gamma'^3)/(\Gamma' - \sin\Gamma'), \tag{9}$$

from (7), with Γ replaced by $\omega_r T$. Hence

$$\omega_r/\omega_e = [1 + F\omega_0 T(\Gamma' - \sin\Gamma')/Q_r\Gamma'^3]^{-1}, \tag{10}$$

or, since the second term dominates,

$$\omega_r/\omega_e \sim (Q_r/Q_m)\times(1-\cos\Gamma')\Gamma'/3(\Gamma'-\sin\Gamma') = (Q_r/Q_m)R(F), \tag{11}$$

in which F is related to Γ' by (8). With a typical residence time of 10^{-4} s, $Q_m \sim 2.5\times 10^6$; it is the large value of Q_m/Q_r that justifies the approximation (11) under normal circumstances. The function $R(F)$ defined by (11) is equal to unity when $F-1$ is small, and falls as F increases, but rather slowly, as shown in fig. 2. With the typical figures quoted, ω_r is of the order of 250 times smaller than ω_e – the molecule, having a much sharper resonance than the resonator, determines the resultant frequency almost entirely, though not so effectively when F is large. If, however, there is plenty of molecular flux in hand Q_r may be reduced, for instance by extracting more of the power from the resonator, so that the maser is only just maintained; as Q_r is thus reduced, the control of the frequency by the molecules is enhanced. It is this very high measure of control that is one factor making the maser a reliable frequency standard. Though good, the ammonia maser is not the best for this purpose, and at the end of this chapter we shall describe an even better system, the hydrogen maser devised by Ramsey. **177**

This analysis also provides a measure of the detuning that is allowable with a given excess flux. Since (8) must have a solution for steady oscillation to occur, the maximum detuning is that which ascribes the whole of Γ' to

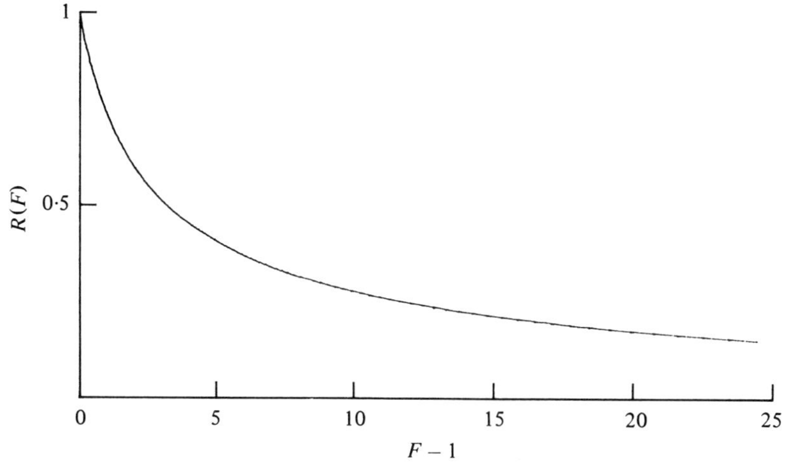

Fig. 2. Variation of $R(F)$ as defined by (11).

Γ, and none to Γ_0;

i.e $$F = (\tfrac{1}{2}\Gamma_{max}/\sin \tfrac{1}{2}\Gamma_{max})^2. \qquad (12)$$

With this unsmoothed model, Γ_{max} cannot exceed 2π, so that $|\omega_r| < \pi/T \sim \omega_0/Q_m$, roughly what we have already inferred by an uncertainty-principle argument.

Fluctuations in amplitude and phase

I.361

In chapter 11 the effect of white noise on a maintained oscillator was shown to be controlled by Q_{sat}, defined in terms of the rate of return to the steady state amplitude after a disturbance; in the case discussed Q_{sat} was the same as the natural Q_r of the unexcited resonant circuit, but this is not always so. In what follows reasonable familiarity with the discussion in chapter 11 will be assumed. The essence of the result derived there is that if the state of the oscillator is represented by a vector defining the amplitude and phase, the end of the vector performs a random walk; but while the excursions of the amplitude are restrained by a radial restoring force, there is no such restraint on tangential motion, and the phase of the oscillation drifts unchecked. Something similar is found in the maser subjected to noise, but the details are different in several important respects. At the root of the difference lies the fact that maser oscillations are maintained by a source that is much more sharply resonant than the system it is exciting, in contrast to the oscillatory circuit in which the amplifier has a very flat frequency response. This makes the analysis rather lengthier – indeed unless the model is carefully devised the mathematical complications are fairly troublesome – and what emerges is that the meaning to be ascribed to Q_{sat} is much closer to Q_m, which is of the order of $\omega_0 T$, than to Q_r: and that white noise does not result in a purely diffusive process for the amplitude and phase.

To reduce the labour of analysis we shall assume that not every molecule enjoys the same residence time T, but that there is an exponential distribution so that after time t only a fraction $e^{-t/T}$ are still reacting coherently to the field; the rest have either left the cavity or have suffered catastrophic collision such as to make them on the average ineffective thereafter. This is in fact a rather good approximation to what happens in the hydrogen maser, and it is in connection with this maser, considered as a frequency standard, that the theory becomes most relevant. We shall concern ourselves only with the case of perfect tuning, when in the representation of fig. 18.1 the electric field $\mathscr{E}_0$ continues to point along η until disturbed by noise. On entering the resonator an excited molecule is represented by a point at the north pole, N, having co-ordinates (0, 0, 1); so long as $\mathscr{E}_0$ is constant and the molecule remains effective the point moves clockwise about the η-axis with angular velocity $p_0\mathscr{E}_0/\hbar$. To find the steady-state value of P_{tot} we need to know $\boldsymbol{R}$, the vector sum of the displacements from the origin of all those molecules that are still effective. Then P_{tot} is p_0 times the projection of $\boldsymbol{R}$ (OP in fig. (3(a)) onto the (ξ, η)-plane. There are three

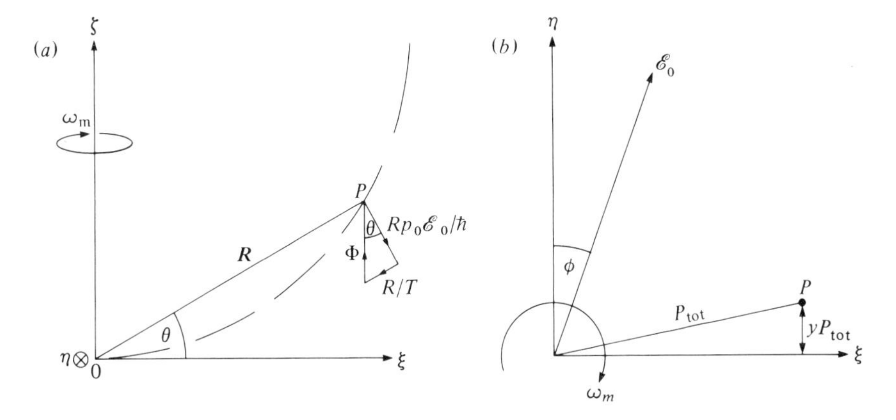

Fig. 3. (a) $\xi-\zeta$ section of fig. 18.1, showing the competing processes which are balanced in the steady state of the double-well. (b) Illustrating the calculation of phase drift due to noise.

processes moving P, which in the steady state must balance – the field causes tangential movement at a speed $Rp_0\mathscr{E}_0/\hbar$: attrition with time-constant T causes movement towards the origin at a speed R/T: and the arrival of new excited molecules causes vertical movement at a speed Φ. These processes are conveniently expressed by treating $\boldsymbol{R}$ as a complex number, writing $R=\xi+i\zeta$. Then

$$\dot{R}=-i(p_0\mathscr{E}_0/\hbar)R-R/T+i\Phi, \tag{13}$$

and in the steady state, when $\dot{R}=0$,

$$R=R_0=i\Phi T/(1+i\Gamma_0), \tag{14}$$

in which Γ_0 has virtually the same significance as before, being defined as $p_0\mathscr{E}_0T/\hbar$. As $\mathscr{E}_0$ is increased, R_0 travels round the circular arc in the diagram, and $\cot\theta=\Gamma_0$.

The total polarization P_{tot} is $p_0 \operatorname{Re}[R_0]$, i.e. $\Phi p_0 T \Gamma_0/(\Gamma_0^2+1)$, and is in phase quadrature with $\mathscr{E}_0$, so that

$$\left.\begin{aligned} &\kappa_e'' = -P_{tot}/\varepsilon_0\mathscr{E}_0 V_{eff} = -\Phi p_0^2 T^2/\hbar\varepsilon_0 V_{eff}(\Gamma_0^2+1) = -F/Q_r(\Gamma_0^2+1), \\ &\text{where} \quad F = \Phi/\Phi_c \text{ as before,} \quad \text{and} \quad \Phi_c = \hbar\varepsilon_0 V_{eff}/Q_r p_0^2 T^2. \end{aligned}\right\} \tag{15}$$

The expression for Φ_c arises from the requirement that at the threshold, when $F=1$ and Γ_0 is small, κ_e'' shall be equal and opposite to the dissipative susceptibility, $1/Q_r$, due to resonator losses. When $F>1$ the steady state has $\Gamma_0^2+1=F$, i.e. $\sin\theta = 1/F^{\frac{1}{2}}$.

Let us return for a moment to vector notation in order to consider the effect of a small arbitrary disturbance in which P is moved from $\boldsymbol{R}_0$ to a point $\boldsymbol{R}_0+\boldsymbol{\rho}$, not necessarily in the same plane. At the same time let $\mathscr{E}_0$ be changed to $\mathscr{E}_0(1+\delta)$ and the phase of oscillation changed by ϕ. As a result $\mathscr{E}_0$ does not now point along the η-axis but has been rotated through ϕ about the ζ-axis. All these perturbations are assumed small enough for linear approximations to hold. The subsequent relaxation of the system back to the steady state can be analysed as two independent processes, movement of P in the plane of $\boldsymbol{R}_0$, and normal to this plane. If $\phi=0$ and $\boldsymbol{\rho}$ lies in the (ξ,ζ)-plane, there is no component of polarization in phase with $\mathscr{E}_0$, so that κ_e' remains zero and the frequency of oscillation is unaffected; consequently ϕ remains at zero and $\boldsymbol{\rho}$ moves only in the plane. Conversely, if $\boldsymbol{\rho}$ has only an η-component, κ_e'' for the total system remains zero and the amplitude of oscillation shows no change; in this case, however, the component of P_{tot} in phase with $\mathscr{E}_0$ introduces a non-vanishing κ_e', so that the frequency is changed and ϕ relaxes to a new steady state. This need not be $\phi=0$, since it is only necessary that the plane containing P shall be normal to $\mathscr{E}_0$ for the frequency to stabilize at ω_0. It is because of this, of course, that noise causes phase-drift.

To take amplitude relaxation first, we once more employ complex notation for the (ξ,ζ)-plane. If $R=R_0+\rho$ and $\mathscr{E}_0$ is changed to $\mathscr{E}_0(1+\delta)$, (13) takes the form, to first order in ρ and δ,

$$\dot{\rho} = -[\rho(1+i\Gamma_0)+i\Gamma_0 R_0\delta]/T. \tag{16}$$

This shows how ρ responds to a given time-variation of $\mathscr{E}_0$, as expressed by $\delta(t)$, and it must be supplemented by an equation showing how $\mathscr{E}_0$ responds to a given time-variation of $p_0 \operatorname{Re}[\rho]$, the change in P_{tot}. Since the steady-state value of P_{tot} suffers a fractional change $\operatorname{Re}[\rho]/\operatorname{Re}[R_0]$, while the field suffers a fractional change δ, the consequent change of the molecular contribution to κ_e'' may be written

$$\Delta\kappa_e''/\kappa_e'' = p_0 \operatorname{Re}[\rho]/P_{tot} - \delta = x - \delta, \tag{17}$$

in which x is the real part of $\rho p_0/P_{tot}$, i.e. of $(\Gamma_0^2+1)\rho/\Phi\Gamma_0 T$. The resonator losses alone would cause the field to decay with time-constant τ_a, a process that would be neutralized by κ_e'' at its steady-state value. According to

(17), however, the time-constant for gain is $\tau_a/(1+\Delta\kappa_e''/\kappa_e'')$, or $\tau_a/(1+x-\delta)$. Consequently

$$\frac{\mathrm{d}}{\mathrm{d}t}[\mathscr{E}_0(1+\delta)] = \mathscr{E}_0(1+\delta)[-1/\tau_a + (1+x-\delta)/\tau_a],$$

so that $$\dot{\delta} = (x-\delta)/\tau_a, \text{ to first order in } \delta \text{ and } x. \quad (18)$$

If z is now written for the imaginary part of $\rho p_0/P_{\text{tot}}$, the real and imaginary parts of (16) provide two further equations to supplement (18):

$$\dot{x} = -(x - \Gamma_0 z - \delta)/T \quad (19)$$

and $$\dot{z} = -(z + \Gamma_0 x + \Gamma_0 \delta)/T. \quad (20)$$

The coupling of x and z in these equations arises from the presence of $1+\mathrm{i}\Gamma_0$ in (16). In consequence, P does not relax back to equilibrium along a straight line but along a spiral path, which may converge quite slowly if F and hence Γ_0 are large; correspondingly the amplitude of $\mathscr{E}_0$ may also show oscillatory relaxation, though the effect is of small importance, as we shall now show.

Equations (18)–(20) yield a third-order equation for δ:

$$\alpha T^3 \dddot{\delta} + (1+2\alpha)T^2\ddot{\delta} + [1+\alpha(\Gamma_0^2+1)]T\dot{\delta} + 2\Gamma_0^2\delta = 0,$$

in which $\alpha = \tau_a/T$ and is very small. The three solutions have the form $\mathrm{e}^{-\lambda t}$, with

$$\lambda \doteqdot [1 \pm \mathrm{i}(8\Gamma_0^2 - 1)^{\frac{1}{2}}]/2T \text{ or } 1/\tau_a. \quad (21)$$

It is the first two solutions that give oscillatory relaxation if $F > 9/8$. These describe a much slower process than the third; (18) shows that δ approaches its momentary steady-state value x with the time-constant τ_a characteristic of the resonator, while (19) and (20) show that x and z themselves are limited by the effective residence time of the molecules to much slower variations. This enables an adequate solution to be constructed without the tedious process of solving the cubic equation exactly. Suppose an impulsive noise-source, acting on the resonator, causes $\mathscr{E}_0$ to be changed slightly, so that at $t=0$, $\delta=\delta_0$ while $x=z=0$. Then in a short interval of a few times τ_a, before x and z have had time to change significantly, δ relaxes to x, i.e. something very close to zero:

$$\delta \doteqdot \delta_0\,\mathrm{e}^{-t/\tau_a}. \quad (22)$$

It is now reasonable to solve (19) and (20) on the assumption that x and z are zero during this interval – i.e. δ provides effectively an impulsive source in these equations, immediately after which $x(+0) = \alpha\delta_0$ and $z(+0) = -\Gamma_0\alpha\delta_0$. Subsequently only the slow solutions in (21) matter, which we match to those initial conditions by writing

$$\left.\begin{aligned} x = \delta &= (\alpha\delta_0 \operatorname{cosec}\varepsilon)\,\mathrm{e}^{-t/2T}\sin(\chi t+\varepsilon) \\ \text{and}\quad \Gamma_0 z &= (\alpha\delta_0 \operatorname{cosec}\varepsilon)\,\mathrm{e}^{-t/2T}[\chi T\cos(\chi T+\varepsilon) - \tfrac{1}{2}\sin(\chi t+\varepsilon)], \end{aligned}\right\} \quad (23)$$

in which χ is the frequency $(8\Gamma_0^2-1)^{\frac{1}{2}}/2T$ given by (21) and $\tan\varepsilon = -2\chi T/(2\Gamma_0^2-1)$.

The variation of δ following an initial impulse is thus made up of a rapidly decaying primary response (22) and a long slow secondary oscillation according to the first equation of (23). The Fourier transform of this impulse response determines the spectrum of the fluctuations of δ when the resonator is subjected to a white noise input. In particular, the low-frequency ($\omega T \ll 1$) components of the fluctuations are simply proportional to the time-integral of the impulse response. On performing the integration, we find that the secondary oscillation changes the integral of (22) alone, which is $\delta_0\tau_a$, to $\delta_0\tau_a(\Gamma_0^2+1)/2\Gamma_0^2$, or $\frac{1}{2}\delta_0\tau_a/(1-1/F)$. When $F>2$ the low-frequency fluctuations are reduced, but never to less than half the amplitude for the passive resonator. So far as amplitude fluctuations are concerned the maser differs little from a maintained oscillator circuit $Q_{\text{sat}} \sim Q_{\text{r}}$.

The same is not true of the phase-drift, which is greatly reduced in the maser. We now assume, as shown in fig. 3(*b*), that P is displaced from the (ξ, ζ)-plane by an amount represented by y so as to give an extra total moment of yP_{tot}. If there were no phase shift of $\mathscr{E}_0$ this would create a real susceptibility κ_{e}' equal to $yP_{\text{tot}}/\varepsilon_0\mathscr{E}_0V_{\text{eff}}$, i.e. y/Q_{r} as follows from (15). If, however, $\mathscr{E}_0$ is advanced in phase by ϕ the component of the moment parallel to $\mathscr{E}_0$ is increased, and

$$\kappa_{\text{e}}' = (y+\phi)/Q_{\text{r}} = 2(y+\phi)/\omega_0\tau_\alpha. \tag{24}$$

There is no first-order change in the component of $\mathscr{E}_0$ normal to the plane of P, and if $\mathscr{E}_0$ were to remain pointing in thc direction ϕ, P would relax back to the plane normal to $\mathscr{E}_0$ with time-constant T, as the molecules grew ineffective and were replaced. Hence

$$\dot{y} = -(y+\phi)/T. \tag{25}$$

On account of (24), however, the resonator frequency is reduced fractionally by $\frac{1}{2}\kappa_{\text{e}}'$, so that

$$\dot{\phi} = -\tfrac{1}{2}\kappa_{\text{e}}'\omega_0 = -(y+\phi)/\tau_a. \tag{26}$$

From (25) and (26),

$$\ddot{\phi} = -\mu\dot{\phi}, \quad \text{where} \quad \mu = 1/\tau_a + 1/T,$$

$$\left.\begin{aligned} &\text{so that} \quad \phi = \phi_1 + \phi_2\,\mathrm{e}^{-\mu t} \\ &\text{and, from (25),} \quad y = -\phi_1 + \alpha\phi_2\,\mathrm{e}^{-\mu t}, \quad \text{where} \quad \alpha = \tau_a/T \text{ as before.} \end{aligned}\right\} \tag{27}$$

If the effect of a noise impulse is to produce an initial phase shift ϕ_0 without immediately changing the polarization, $\phi(0) = \phi_0$ and $y(0) = 0$. Hence $\phi_1 + \phi_2 = \phi_0$ and $\phi_1 = \alpha\phi_2$;

$$\text{i.e.} \qquad \phi = \phi_0(\mathrm{e}^{-\mu t} + \alpha)/(1+\alpha).$$

The immediate phase shift is rapidly undone by the molecules; after a few times τ_a it is reduced to $\alpha\phi_0$, if $\alpha \ll 1$. Now according to (11.44) the Gaussian spread of phase in time t, for a given voltage noise, is inversely proportional to the circuit inductance, and hence to Q_r (since the given noise source implies R is constant). Thus the result we have derived implies that the phase drift is controlled by the molecular resonance, having $Q_m = \omega_0 T$, not the resonator with $Q_r = \omega_0 \tau_e$. It is as if, roughly speaking, the resonator and the excited molecules were to be treated as a single system, with the latter contributing so much energy that the total energy of oscillation (as may easily be shown) is something like Q_m/Q_r times as great as the field energy alone; the losses and the noise input, however, are due to the resonator only and the combined system therefore behaves like a simple oscillatory circuit with decay time T, and correspondingly enhanced phase stability.

Random noise input is not the only source of fluctuations; statistical variations in the flux of molecules give rise to amplitude variations but not to phase drift in a well-tuned system. This process can be shown to simulate thermal noise corresponding to a temperature of about $\hbar\omega_0/k_B$, which is some hundreds of times less than room temperature for the ammonia maser. The effect is important in optical masers, where $\hbar\omega_0 \gg k_B T$, but we shall not analyse it here. In addition there is, as already mentioned, a purely quantal source of noise which the analysis in terms of dielectric response misses altogether; it operates on the phase as well as the amplitude to cause fluctuations in both. We shall find it emerging automatically, if only as an incidental feature, from a fairly rigorous quantal treatment which goes a long way, apart from this point, towards justifying the dielectric approach. It is reasonable to expect that each process will be buffered by the molecules in the same way as thermal noise. The molecules do not interact directly with one another, but only indirectly through their effect on the resonator field. Any perturbation of phase by one molecule, therefore, whether of quantum or statistical origin, first changes $\mathscr{E}$ and the change is then diminished by a factor τ_a/T as the other molecules react.

Quantum mechanics of the resonator–molecule interaction

Once a molecule has come into interaction with the resonator the wave-functions of the two remain inseparable until an observation of the state of the resonator enables a new start to be made. We are interested in the probable state of the resonator after the maser has been running freely for a time, and must therefore keep in the composite wave-function all the molecules that have passed through during that time. This proves a far less formidable proceeding than might be feared since the molecules, after leaving the resonator, are effectively uncoupled from it and from each other, and the stationary states of an assembly of uncoupled systems are represented simply as products of eigenfunctions of individual members of the assembly. We shall permit ourselves one simplification which certainly does not accord with the facts, by assuming only one molecule to be in interaction at any instant, the total effect on the resonator being the sum of all such binary interactions. In the light of the foregoing discussion, and somewhat in the spirit of the Hartree approximation, we may suppose that the cavity and all molecules except the one under consideration are so closely coupled as to behave as a single resonator, with decrement governed by T rather than τ_e. Thus we shall work out the simplified model on the assumption that only one molecule is actually in the resonator cavity at any time, but shall then replace Q_r by Q_m when it seems reasonable. Similarly, although the empty resonator may be mistuned by ω_e with respect to the molecules, ultimately this mistuning is reduced to ω_r by the combined action of all molecules present, and we shall assume that ω_r is the mistuning experienced by any one molecule during its passage. It is perhaps distasteful to have recourse to such adjustments during what pretends to be a rather rigorous treatment, but the alternative involves so much more analysis that the basic physics may well be lost. By the procedure adopted here the quantum mechanics is at all events reduced to a fairly simple operation – we follow one molecule through the resonator to see how the many-body wave-function is thereby changed. In particular we determine the evolution of $\mathcal{P}(x, t)$, the probability that at time t the harmonic oscillator representing the resonator has displacement x. We have no interest in the molecules as such, but only in the state of the resonator mode as a result of their interactions with it. From now on it will be assumed that each molecule spends the same time T in the resonator rather than treating the exponential distribution of residence times as in the last section.

The analysis differs in detail according to the state of the molecule at the moment of entry; we shall work through the analysis for an excited molecule and at the end indicate what changes are needed for one that is unexcited. Just before the new molecule enters the resonator the wave-function of the entire system can be expressed as a Fourier sum over many-body eigenfunctions, each of which, since all parts are uncoupled, is a product of individual eigenfunctions, e.g. an oscillator function $\psi_n(x)$, a multitude of two-level functions $\chi_{1,2}(q_i)$ for previous molecules, and the initial wave-function of the new molecule $\chi_2(y)$. The $\chi(q_i)$ remain uncoupled, and their product may be gathered under a single umbrella function $U_m(q_i)$. Then since the same $\chi_2(y)$ appears in all eigenfunctions making up the initial state we write for the wave-function of the whole at the outset $(t=0)$:

$$\Psi(x, y, q_i; 0) = \chi_2 \sum_n \sum_m A_{nm} \psi_n U_m, \tag{28}$$

all possibilities being encompassed in the choice of A_{nm}. Once the new molecule is coupled to the resonator, $\chi_2\psi_n$ ceases to be an eigenfunction for thc two and must be rewritten as a sum of the true eigenfunctions, as given by (18.33):

$$\Psi(x, y, q_i; 0) = \sum_n \sum_m A_{nm} U_m [\beta_n(\alpha_n \chi_1 \psi_{n+1} + \beta_n \chi_2 \psi_n) - \alpha_n(\beta_n \chi_1 \psi_{n+1} - \alpha_n \chi_2 \psi_n)], \tag{29}$$

in which α_n and β_n are given by (18.38) and both are positive. The two terms in (29) now oscillate at different frequencies. No time-variation that is common to all the eigenfunctions in (29) plays any role; we may therefore ignore the oscillations associated with U_m and take the two terms in square brackets to have frequencies $(n+1)\omega \pm \Omega_n$, where Ω_n is given by (18.36) with ω_e replaced by ω_r. Also it should be noted that the resultant frequency, ω, is used instead of ω_0.

Then
$$\Omega_n = [\omega_r^2 + \gamma^2(n+1)]^{\frac{1}{2}}, \tag{30}$$

and Ω_n has the same sign as ω_r. To find how Ψ develops, we must multiply the first term in (29) by $\exp i[-(n+1)\omega - \Omega_n]t$ and the second term by $\exp i[-(n+1)\omega + \Omega_n]t$. Then after some rearrangement

$$\Psi(x, y, q_i; t) = e^{-i\omega t} \sum_n \sum_m U_m \psi_n [A_{nm}\chi_2(\beta_n^2 e^{-i\Omega_n t} + \alpha_n^2 e^{i\Omega_n t}) - 2iA_{n-1,m}\alpha_{n-1}\beta_{n-1}\chi_1 \sin \Omega_{n-1}t] e^{-in\omega t}. \tag{31}$$

Now
$$\mathscr{P}(x, t) = \int \cdots \int \Psi^*\Psi \, dq_i \, dy, \tag{32}$$

and if Ψ^* is written in the same form as (31), but with subscripts k, l instead of n, m, the integrals over q_i are separable in the form $\int \cdots \int U_k^* U_m \, dq_i = \delta_{km}$. Hence the U's disappear from (32), which now becomes:

$$\mathscr{P}(x, t) = \sum_l \sum_m \sum_n e^{i(l-n)\omega t} \int \psi_l^* \psi_n \, dy [A_{lm}^* \chi_2^* e^{i\omega t}(\beta_l^2 e^{i\Omega_l t} + \alpha_l^2 e^{-i\Omega_l t}) + 2iA_{l-1,m}^* \alpha_{l-1}\beta_{l-1}\chi_1^* \sin \Omega_{l-1}t][A_{nm}\chi_2 e^{-i\omega t}(\beta_n^2 e^{-i\Omega_n t} + \alpha_n^2 e^{i\Omega_n t}) - 2iA_{n-1,m}\alpha_{n-1}\beta_{n-1}\chi_1 \sin \Omega_{n-1}t]. \tag{33}$$

Since χ_1 and χ_2 are orthogonal and normalized, integration over y is trivial, and

$$\mathscr{P}(x, t) = \sum_l \sum_n e^{i(l-n)\omega t} \psi_l^* \psi_n [\rho_{ln}(\beta_l^2 e^{i\Omega_l t} + \alpha_l^2 e^{-i\Omega_l t})(\beta_n^2 e^{-i\Omega_n t} + \alpha_n^2 e^{i\Omega_n t}) + 4\rho_{l-1,n-1}\alpha_{l-1}\beta_{l-1}\alpha_{n-1}\beta_{n-1} \sin \Omega_{l-1}t \sin \Omega_{n-1}t], \tag{34}$$

in which ρ_{ln} is written for the density matrix:

$$\rho_{ln} = \sum_m A_{lm}^* A_{nm} = \rho_{nl}^*. \tag{35}$$

If the original wave-functions were normalized, $\Sigma_n \rho_{nn} = 1$ automatically. This concludes the quantum-mechanical analysis, which we now proceed to interpret.

At the moment the molecule enters ($t = 0$) let us write ρ_{ln} as $\bar{\rho}_{ln}$; then

$$\mathscr{P}(x, 0) = \sum_l \sum_n \bar{\rho}_{ln} \psi_l^* \psi_n, \tag{36}$$

and at the moment it leaves ($t = T$) (34) shows that

$$\mathscr{P}(x, t) = \sum_l \sum_n e^{i(l-n)\omega T} \rho_{ln}^{+} \psi_l^* \psi_n, \tag{37}$$

in which

$$\left.\begin{aligned} \rho_{ln}^{+} &= (1+\lambda_{ln})\rho_{ln}^{-} + \mu_{l-1,n-1}\rho_{l-1,n-1}^{-}, \\ \lambda_{ln} &= (\beta_l^2\, e^{i\Omega_l T} + \alpha_l^2\, e^{-i\Omega_l T})(\beta_n^2\, e^{-i\Omega_n T} + \alpha_n^2\, e^{i\Omega_n T}) - 1, \\ \text{and} \quad \mu_{ln} &= 4\alpha_l\beta_l\alpha_n\beta_n \sin \Omega_l T \sin \Omega_n T \end{aligned}\right\} \tag{38}$$

During the passage of one molecule the phase of each term in (37) advances at its characteristic rate, $(l-n)\omega$, while ρ_{ln}^{-} is changed to ρ_{ln}^{+}. If no molecule is passing through, ρ_{ln} remains unchanged but the phase advance continues unabated. We may therefore proceed to the effect of a steady stream of molecules by writing

$$\mathscr{P}(x, t) = \sum_l \sum_n e^{i(l-n)\omega t} \rho_{ln}(t) \psi_l^* \psi_n, \tag{39}$$

and using (38) to define the change suffered by ρ_{ln} through interaction with a single excited molecule:

$$\delta\rho_{ln} = \lambda_{ln}\rho_{ln} + \mu_{l-1,n-1}\rho_{l-1,n-1}. \tag{40}$$

For a flux of Φ molecules in unit time,

$$\dot{\rho}_{ln} = \Phi(\lambda_{ln}\rho_{ln} + \mu_{l-1,n-1}\rho_{l-1,n-1}). \tag{41}$$

If the elements of ρ_{ln} are set out in a square array, each line of elements parallel to the diagonal develops independently, according to (41). On the diagonal itself, where $l = n$, (38) shows that

$$-\lambda_{nn} = \mu_{nn} = 4\alpha_n^2\beta_n^2 \sin^2 \Omega_n T. \tag{42}$$

Hence $\Sigma_n \dot{\rho}_{nn} = 0$, and the trace of the matrix, $\Sigma_n \rho_{nn}$, is independent of time, as already noted; the elements form a conserved distribution along the diagonal which may evolve in form but certainly climbs steadily to higher n as each ρ_{nn}, according to (41), acquires some of the distribution lying next below it and passes on its own distribution to the level above. Since each molecule in its passage moves ρ_{nn} upwards by a fraction μ_{nn} of an integer, μ_{nn} may be interpreted as the probability that a molecule emerges unexcited.

The off-diagonal lines do not conserve their distribution but suffer gradual dissipation. To show this it is enough to expand $\lambda_{ln} + \mu_{ln}$ as a power series in $n - l(\equiv \nu)$, stopping at the leading real and imaginary terms. After some tedious evaluations involving the derivatives of (38) and (30) we find

$$\sum_n \dot{\rho}_{n-\nu,n} = \Phi \sum (\lambda_{n-\nu,n} + \mu_{n-\nu,n})\rho_{n-\nu,n},$$

and

$$\begin{aligned} \lambda_{n-\nu,n} + \mu_{n-\nu,n} \approx &-(\gamma^4 T^2 \nu^2/8\Omega_n^2)[1 + (\omega_r^2/\gamma^2 n \Omega_n^2 T^2) \sin^2 \Omega_n T] \\ &+ i(\gamma^2 \omega_r \nu/4\Omega_n^3)[2\Omega_n T - \sin 2\Omega_n T]. \end{aligned} \tag{43}$$

Only on the diagonal, $\nu = 0$, does $\Sigma_n \dot{\rho}_{n-\nu,n}$ vanish; elsewhere it is negative, and the sum decays at a rate that depends on n and is also proportional to ν^2. We shall defer discussion of this and of the frequency shift that results from the imaginary part of (43).

The equation (41) describing the development of the density matrix was worked out for the case of excited molecules entering the resonator. If instead we had assumed unexcited molecules the step from (28) to (29) would have been different, with χ_1 instead of χ_2 in (28) and correspondingly $[\alpha_{n-1}(\alpha_{n-1}\chi_1\psi_n + \beta_{n-1}\chi_2\psi_{n-1}) + \beta_{n-1}(\beta_{n-1}\chi_1\psi_n - \alpha_{n-1}\chi_2\psi_{n-1})]$ in (29); the two eigenfunctions in the square brackets have frequencies $n\omega \pm \Omega_{n-1}$. As a result (38) must be rewritten in the form

$$\left.\begin{aligned} \rho^+_{ln} &= (1+\lambda'_{ln})\rho^-_{ln} + \mu'_{l+1,m+1}\rho^-_{l+1,n+1}, \\ \text{where} \qquad \lambda'_{ln} &= \lambda^*_{l-1,n-1} \quad \text{and} \quad \mu'_{l+1,n+1} = \mu_{ln}. \end{aligned}\right\} \tag{44}$$

Hence (41) now has the form

$$\dot{\rho}_{ln} = \Phi(\lambda'_{ln}\rho_{ln} + \mu'_{l+1,n+1}\rho_{l+1,n+1}), \tag{45}$$

which describes a steady progression of the distribution downwards towards lower l, n.

Graphical representation of density matrix

21

The σ-representation, the ensemble of classical oscillators that was introduced in chapter 13, is readily extended to describe the more versatile function $\mathcal{P}(x, t)$ in (39). If the cavity is left undisturbed, so that each ρ_{ln} is constant, the probability $\mathcal{P}(x, t)$ of finding it displaced by x at time t is made up of the superposition of sinusoidally oscillating real functions, of which $\rho_{n-\nu,n}\,\mathrm{e}^{\mathrm{i}\nu\omega t}\psi^*_{n-\nu}\psi_n + \rho_{n,n-\nu}\,\mathrm{e}^{-\mathrm{i}\nu\omega t}\psi^*_n\psi_{n-\nu}$ is a typical example, the two terms being complex conjugates by virtue of (35). This represents a certain spatial variation of $\mathcal{P}$ with x which oscillates harmonically at a frequency of $\nu\omega$. It is always possible to find a real radial density $\sigma(r)$ such that the two-dimensional distribution $\sigma(r)\cos(\nu\theta + \varepsilon)$ when spun at angular velocity ω projects, as explained in chapter 13, onto the x-axis the desired $\mathcal{P}(x)$ oscillating at a frequency $\nu\omega$.† Therefore any $\mathcal{P}(x, t)$ described by ρ_{ln} has its counterpart in the form of an ensemble of classical oscillators of which the number displaced by an amount x at time t is proportional to $\mathcal{P}(x, t)$. The distribution $\sigma(r, \theta)$ defines the number of oscillators at any instant with amplitude r and phase θ, and in an undisturbed system $\sigma(r, \theta)$ spins unchanged. The diagonal terms of the density matrix are represented by axially symmetric forms of $\sigma(r, \theta) = \sigma_0(r)$, while all terms on the off-diagonal lines defined by $l - n = \pm\nu$ are represented collectively by $\sigma_\nu(r)\cos(\nu\theta + \varepsilon_\nu)$.

† If $\sigma(r)$ is the Bessel (Hankel)$^{(2)}$ function $H_\nu(kr)$, the projection of $\sigma(r)$ has the form $\mathrm{e}^{\mathrm{i}kx}$. Any periodic $\mathcal{P}(x, t)$ may therefore be Fourier-analysed in x and t, and the amplitude at frequency $\nu\omega$ and wavenumber k used as the amplitude of the corresponding $H_\nu(kr)$ to synthesize the required $\sigma(r)$.

The coherent state (13.16) and (13.18), represented by a narrowly confined hump of σ running around a ring, requires a certain spread in the values of n to provide tangential confinement. We know, in fact, from (13.23), that the wave-function of the isolated oscillator in the coherent state centred on n_c contains each ψ_n with amplitude a_n roughly proportional to $e^{-(n-n_c)^2/4n_c}$, and from the definition of A_{nm} in (28) it is clear that m is now irrelevant and that A_{nm} is simply the amplitude of ψ_n. The density matrix is then a product of the separate amplitudes,

12

$$\rho_{ln} = \alpha_l^* a_n \propto e^{-\Delta^2/2n_c} \tag{46}$$

where $\Delta^2 = (l-n_c)^2 + (n-n_c)^2$. The coherent state appears as an axially symmetrical Gaussian hump in both the equivalent classical ensemble and the density matrix.

The restriction of ρ_{ln} to a product, as in (46), is of course not peculiar to the coherent state, but is characteristic of any pure state, that is to say, one in which the behaviour of the oscillator is expressible in terms of oscillator eigenfunctions alone. When the oscillator is coupled to other systems it is in a mixed state, necessitating eigenfunctions of all the systems involved to describe it, and it is then that the more general form (35) for ρ_{ln} makes its appearance. It is clear from the dissipative property of (43) that as time goes on the coupling inevitably results in ρ_{ln} becoming more strongly confined to the diagonal (in which form it cannot be expressed as a product). Correspondingly in the σ-representation much of the tangential detail fades away until in the end, when ρ_{ln} is purely diagonal, σ is axially symmetric, though it may still be strongly peaked around a certain radius r. Such a distribution, in which the oscillator may have a well-defined amplitude (or energy) but no preference as to phase, cannot be expressed by any combination of oscillator wave-functions, but is a characteristic example of the greater generality allowed by the density matrix. It is interesting to observe how the coupling of quantum systems leads to new effects such as this, without immediately ruling out the use of a classical model to represent the behaviour. We shall even find that the rules governing the development of the model are visualizable in classical terms. Let us derive these rules.

The dissipative harmonic oscillator

The essential results can be demonstrated without stimulating the resonator into self-sustained maser action, and at the same time the earlier, rather sketchy, discussions of dissipative processes can be rounded out. It was pointed out by Scully and Lamb[(3)] that a stream of unexcited molecules passing through a resonator served to damp its oscillation, and that this mechanism was considerably easier to work out completely than damping by radiation, since each element of the dissipative process is a two-level system rather than a cavity mode with its ladder of equally spaced levels. There is no need for the molecules to be well tuned to the resonator, since

75, 104

even if poorly tuned they still have a chance, albeit small, of emerging excited.† It is positively advantageous, indeed, to model the dissipation with a very high flux Φ_d of very weakly coupled molecules, since the fluctuations in the flux can thereby be made negligibly small. All random effects can then be attributed to the interaction of the quantized systems and, as we shall see, it is easy to include thermal noise as well. Moreover, if we assume the residence time T to be much shorter than τ_e, all the problems discussed earlier about the appropriate value of Q disappear. We suppose that there are equal fluxes of molecules with positive and negative mistuning, ω_r, so that the imaginary part of (43) disappears. In the real part the second term in the square brackets can be made large by reducing the coupling constant γ, so that only this term need be retained.

† As described here, the process appears not to conserve energy – an unexcited molecule, whose frequency ω_m is not the same as the cavity frequency ω_0, may gain $\hbar\omega_m$ while the cavity loses $\hbar\omega_0$. It is not enough to take refuge in the Uncertainty Principle with the observation that a transition is only likely if $|\omega_m - \omega_0|T < 1$; this may be true, but the fact remains that the initial and final states of both molecule and cavity can be ascertained at leisure, with arbitrarily small uncertainty. To resolve the problem it is necessary to bring in the translational kinetic energy of the molecule, which does not remain constant during the interaction. The physical mechanism that causes it to change may be traced to the fringing field at the entrance and exit of the cavity; as the molecule crosses the threshold it begins to be polarized, and the gradient of field acts on the dipole moment to exert a translational force. In the light of this let us see how a stationary-state wave-function might be constructed to describe the complete process, involving the cavity and the whole path of the molecule from before its entry until after its exit. Initially we might have the molecule and cavity in well-defined states, $\chi_1 \mathrm{e}^{\mathrm{i}k_0x}$ for the molecule and ψ_m for the cavity; the term $\mathrm{e}^{\mathrm{i}k_0x}$ describes the initial translational motion. If the total energy, including kinetic energy, is E, the complete wave-function is $(\psi_n\chi_1 \mathrm{e}^{\mathrm{i}k_0x})\,\mathrm{e}^{-\mathrm{i}Et/\hbar}$. This must be matched, at the entrance to the cavity, to the wave-function inside, where because of the interaction the relevant eigenstates will have the form $\alpha\chi_1\psi_n + \beta\chi_2\psi_{n-1}$ as in (18.33), with two choices of α and β. The complete wave-function will be a superposition of the two eigenstates, each multiplied by its appropriate $\mathrm{e}^{\mathrm{i}k_1x}$ or $\mathrm{e}^{\mathrm{i}k_2x}$ so as to give the same total energy E; only then can the matching persist at all times. The 'internal' energy difference $2\hbar\Omega_n$ is compensated by the difference in kinetic energy $\hbar^2(k_1^2 - k_2^2)/2m$, or $\hbar^2 k_0\delta k/m$ if the coupling is weak. Because of the wavenumber difference, at the exit the two waves will have suffered a relative phase shift of $L\delta k$, and will not recombine outside to form the original wave-function once more. Instead there will be a mixture of the original $(\psi_n\chi_1 \mathrm{e}^{\mathrm{i}k_0x})\,\mathrm{e}^{-\mathrm{i}Et/\hbar}$ and $(\chi_{n-1}\chi_2 \mathrm{e}^{\mathrm{i}k_0'x})\,\mathrm{e}^{-\mathrm{i}Et/\hbar}$, with k_0' different from k_0 so as to maintain the same total energy E. Clearly, if an electrostatic lens is used to separate these two states it will be found that whichever path any given molecule takes it will emerge with exactly the right energy. In summary, when in the text it is argued that the two eigenfunctions in the cavity suffer a phase shift of $2\Omega_n T$ because their energies differ, we now attribute the resulting transition probability to the difference in wavenumber, and write the phase shift as $L\delta k$, i.e. $2\Omega_n mL/\hbar k_0$. Since the molecular speed is $\hbar k_0/m$, these two expressions are equivalent. We may therefore continue to use the argument in the text, secure in the knowledge that energy is conserved in spite of appearances.

Then, since $\sin(\Omega_n T) \approx \Omega_n T$ when T is small, and $\Omega_n \approx \omega_r$ for large mistuning,

$$\lambda'_{n-\nu,n} + \mu'_{n-\nu,n} \approx -\gamma_d^2 T_d^2 \nu^2 / 8n, \tag{47}$$

in which γ_d and T_d are used to distinguish the coupling constant and residence time of the dissipative molecules from those of the molecules responsible for maser amplification. When $\nu = 0$, on the diagonal,

$$\mu'_{nn} = -\lambda'_{nn} = 4\alpha_{n-1}^2 \beta_{n-1}^2 \sin^2(\omega_r T_d), \text{ from (38) and (44)},$$

$$\approx \gamma_d^2 T_d^2 n, \text{ from (18.38) and (18.36).} \tag{48}$$

If we write B_d for $\Phi_d \delta_d^2 T_d^2$, the development of ρ_{ln} may be summarized thus. On the diagonal, from (45) and (48),

$$\dot{\rho}_{nn} = B_d[-n\rho_{nn} + (n+1)\rho_{n+1,n+1}]. \tag{49}$$

Off the diagonal, $-\lambda_{n-\nu,n}$ and $\mu_{n-\nu,n}$ are almost the same as when $\nu = 0$, so that (49) very nearly describes the development of ρ_{ln}. However, the difference between μ' and $-\lambda'$, which is smaller than either by something like ν^2/n^2, leads to non-conservation since, from (45),

$$\sum_n \dot{\rho}_{n-\nu,n} = \Phi_d \sum_n (\lambda'_{n-\nu,n} + \mu'_{n-\nu,n})\rho_{n-\nu,\nu} = -\tfrac{1}{8}\mathrm{B}_d \nu^2 \sum_n \rho_{n-\nu,n}/n. \tag{50}$$

A reasonably narrow distribution, of width $n^{\frac{1}{2}}$ say, allows n to be taken outside the summation in (50);

$$\sum_n \dot{\rho}_{n-\nu,n} \approx -(\mathrm{B}_d \nu^2 / 8\bar{n}) \sum_n \rho_{n-\nu,n}, \tag{51}$$

indicating that while this line of the matrix develops on the whole in the same way as the diagonal, the distribution decays steadily, but not exponentially since the instantaneous time-constant varies as $\bar{n}$, which is steadily falling.

We discuss the two aspects of the development separately; first the diagonal terms and then the decay of the off-diagonal terms. Since (49) is linear in ρ_{nn}, any solution may be synthesized once we know how a single term develops. The appropriate Green's function has the form

$$\rho_{rr}(t') = {}^N\mathrm{C}_r (\mathrm{e}^{t'} - 1)^{N-r}\, \mathrm{e}^{-Nt'}, \tag{52}$$

which will be found by substitution to satisfy (49) if $t' = B_d t$. To find the initial state, let $t' \to 0$; then $\rho_{rr} \to 0$ for all r except $r = N$, for which $\rho_{NN} = 1$. The subsequent conservation of $\Sigma_r \rho_{rr}$ follows by considering the generating function $(1 - \mathrm{e}^{-t'})^N [1 + 1/(\mathrm{e}^{t'} - 1)]^N$; on expanding the square brackets as a binomial series, the rth term is seen to be ρ_{rr}, and since the function itself is obviously equal to unity it follows that $\Sigma_r \rho_{rr} = 1$. A convenient approximation to (52) results from applying Stirling's approximation to (52) followed by Taylor expansion:

$$\rho_{rr} \sim \mathcal{N}_r\, \mathrm{e}^{-N(r-r_m)^2/2r_m(N-r_m)}, \tag{53}$$

in which r_m, which defines the position of the peak, is $N\,\mathrm{e}^{-t'}$, and $\mathcal{N}_r$ is a time-dependent normalizing coefficient to conserve ρ_{rr}. The decay of r_m, and hence of energy, as $e^{-t'}$ shows that $t = t/\tau_e$, i.e. $B_d = 1/\tau_e$. The mean square width of the Gaussian peak follows from (53):

$$\overline{(r - r_m)^2} = r_m(1 - r_m/N) = N\,e^{-t'}(1 - e^{-t'}). \tag{54}$$

As the centre of the peak falls exponentially with time from N to zero, the R.M.S. width expands from zero to a maximum of $\frac{1}{2}N^{\frac{1}{2}}$, when $r_m = \frac{1}{2}N$, and then decreases, ultimately as $r_m^{\frac{1}{2}}$. Stirling's approximation renders (54) unreliable when r_m is small, but (52) shows that, as expected, the system ends up in the ground state.

We have just determined the development of the density matrix starting from a pure eigenstate, ψ_N. If we had started with a diagonal ρ_{nn} in the form of a narrow Gaussian peak centred on N and of mean square width w_0, $\rho_{nn} \propto e^{-(n-N)^2/2w_0}$, each element of ρ_{nn} would have developed according to (53), so that after time t the distribution of centres would have collapsed to a peak of mean square width $w_0\,e^{-2t'}$ centred on $N\,e^{-t'}$. On the other hand, each element would now have spread into a peak of width given by (54), so that the resultant would be the convolution of two Gaussian curves; such a convolution is itself Gaussian, with a mean square width w that is the sum of the separate mean square widths:

$$w = w_0\,e^{-2t'} + N\,e^{-t'}(1 - e^{-t'}). \tag{55}$$

In particular, if we start with $w_0 = N$, subsequently $w = N\,e^{-t'} = r_m$. This means that the dissipative decay of a distribution that has been set up to match the coherent state of the oscillator, with a mean square width of the n-distribution equal to the mean value of n itself, maintains the coherent state throughout.

This result was derived for the diagonal terms of ρ_{ln}, but it holds for the off-diagonal terms also. To verify this point, let us start with the coherent pure state in which ρ_{ln} is an isotropic Gaussian peak of mean square width N centred on the point (N, N),

$$\rho_{ln} \propto e^{-[(l-N)^2+(n-N)^2]/4N}. \tag{56}$$

Along any off-diagonal line $l = n - \nu$, the distribution is Gaussian, with the same width N, and height reduced by $e^{-\nu^2/8N}$. Now let this whole peak move down towards the origin, remaining isotropic while shrinking in width so that the mean square width is always equal to the value of n at the centre. When the centre has reached $\bar{n}$ the off-diagonal section has a peak height reduced from that at the centre by a factor $e^{-\nu^2/8\bar{n}}$. In order to maintain the coherent state the off-diagonal elements must develop in the same way as the diagonal elements, except that they must decay at the same time. Thus the sum of ρ_{ln} along a line of constant ν, which is constant if $\nu = 0$, must in general have the property that

$$\sum_n \rho_{n-\nu,n} \propto e^{-\nu^2/8\bar{n}}.$$

Then, by taking the logarithmic derivative, we have that

$$\sum_n \dot{\rho}_{n-\nu,n} = (\nu^2/8\bar{n}^2)\dot{\bar{n}} \sum_n \rho_{n-\nu,n} = (-\nu^2/8\bar{n}\tau_e) \sum_n \rho_{n-\nu,n}, \tag{57}$$

since $\bar{n}$ decays exponentially with time-constant τ_e. Now we have already found that $B_d = 1/\tau_e$, so that (51) and (57) are identical.

We have demonstrated that when an oscillator in a coherent state suffers dissipation, it remains in a coherent state. This result is immediately translatable into the σ-representation, since we already know what distribution $\sigma(\boldsymbol{R})$ describes the coherent state; it has the form of the ground state distribution e^{-R^2}, where $\boldsymbol{R}$ is the vector (ξ, κ), but displaced from the origin to a new centre, $\boldsymbol{R}_0$ say; $\sigma \propto \exp\{-|\boldsymbol{R}-\boldsymbol{R}_0|^2\}$. The result just derived shows that if we allow this coherent state to decay the distribution remains unchanged in form and size while its centre $\boldsymbol{R}_0$ relaxes exponentially back to the origin. The whole pattern is of course spinning at angular frequency ω_0 round the origin, but we take this for granted. A relaxation of this character is characteristic of a distribution of points which are subject to an organized drift towards the origin, at a speed proportional to displacement, superimposed on diffusive motion obeying Fick's law. If the concentration of points is $\sigma(\boldsymbol{R})$, the local flux $\boldsymbol{V}$, as determined by these processes, is $-\alpha\sigma\boldsymbol{R} - (D \operatorname{grad} \sigma)$. The first term is due to organized drift and the second to diffusion. In the steady state $\sigma \propto e^{-\alpha R^2/2D}$, for which distribution $\boldsymbol{V} = 0$ everywhere. Let us now shift this pattern to a new centre, $\boldsymbol{R}_0$. At any other point $\boldsymbol{R}$ the organized drift velocity is $-\alpha\boldsymbol{R}$ while the diffusive velocity is away from $\boldsymbol{R}_0$ and of magnitude $\alpha(\boldsymbol{R}-\boldsymbol{R}_0)$. The resultant $\boldsymbol{V} = -\alpha\sigma\boldsymbol{R}_0$, so that every part of the σ-distribution moves at the same speed, $\alpha\boldsymbol{R}_0$. If there was no random element associated with the dissipation the distribution would diminish in radius as every part relaxed towards the origin; but the random effect of the dissipative mechanism, represented by the diffusive term, counteracts this diminution. The angular spread of the distribution, i.e. the uncertainty in phase of the coherent oscillation, may be very small, say $\bar{n}^{-\frac{1}{2}}$, when $\bar{n}$ is large. Clearly, however, it increases as dissipation drives the distribution towards the origin; eventually the distribution is isotropic round the origin, and no phase information remains. This is a simple pictorial representation of the decay of the off-diagonal elements of ρ_{ln}.

We have exhibited a special case of the diffusive effect, but it can be proved quite generally by returning to ρ_{ln} and remembering that every line $l = n - \nu$ develops independently of the others. What follows is an outline of the argument, without details of the manipulations. Considering one such line whose contribution to $\mathscr{P}(\xi, t)$ may be written, following (39),

$$\mathscr{P}_\nu(\xi, t) = \sum_n \rho_{n-\nu,n} \Psi^*_{n-\nu} \Psi_n, \tag{58}$$

we take the ψ_n as real and ignore the oscillatory term, which is the same as ignoring the spinning of the σ-distribution. It is convenient to write ψ_n

in terms of Hermite polynomials, as in (13.3):

$$\psi_n = \pi^{-\frac{1}{4}}(2^n n!)^{-\frac{1}{2}} H_n(\xi)\, e^{-\frac{1}{2}\xi^2},$$

and to make use of the following properties of the polynomials:[4]

$$\left.\begin{aligned} &\mathrm{d}H_n/\mathrm{d}\xi = 2nH_{n-1} \\ \text{and} \qquad &\mathrm{d}^2H_n/\mathrm{d}\xi^2 = 4n\xi H_{n-1} - 2nH_n. \end{aligned}\right\} \qquad (59)$$

From (58) and (45) we have that

$$\partial\mathcal{P}_\nu/\partial t - \Phi_\mathrm{d} \sum_n \rho_{n-\nu,n}(\lambda'_{n-\nu,n}\psi_{n-\nu}\psi_n + \mu'_{n-\nu,n}\psi_{n-\nu-1}\psi_{n-1}).$$

In addition, the argument leading to (48) may be extended to show that

$$\Phi_\mathrm{d}\lambda'_{n-\nu,\nu} = -(n-\tfrac{1}{2}\nu)B_\mathrm{d} = -(n-\tfrac{1}{2}\nu)/\tau_\mathrm{e} \quad \text{and} \quad \Phi_\mathrm{d}\mu'_{n-\nu,\nu} = [n(n-\nu)]^{\frac{1}{2}}/\tau_\mathrm{e}.$$

Given this information, the reader may verify that

$$\partial^2\mathcal{P}_\nu/\partial\xi^2 + 2\partial(\mathcal{P}_\nu\xi)/\partial\xi = 4\tau_\mathrm{e}\partial\mathcal{P}_\nu/\delta t, \qquad (60)$$

independent of ν and therefore true for any $\mathcal{P}(\xi, t)$.

Since this equation continues to hold as the σ-distribution turns round, and since $\mathcal{P}$ is the projection of σ onto the ξ-axis, (60) must represent one component of the development of σ itself, so that we may write

$$D\nabla^2\sigma + 2D\,\mathrm{div}\,(\sigma R) = \partial\sigma/\mathrm{d}t, \qquad (61)$$

where $D = 1/4\tau_\mathrm{e} = \omega_0/4Q_\mathrm{r}$. The left-hand side is the divergence of a flux vector $-D\,\mathrm{grad}\,\sigma + \sigma\boldsymbol{R}/2\tau_\mathrm{e}$, in which the second term describes exponential decay of the individual vectors with a time-constant $2\tau_\mathrm{e}$, i.e. τ_a.

This result looks remarkably classical. If $\sigma(\boldsymbol{R})$ is considered as representing an ensemble of real classical vibrators, subject to linear dissipative forces and to random noise, the latter causes each point in the distribution to execute a random walk which is described statistically by a diffusion coefficient. Such an equation of motion for σ is identical with (61), and the magnitude of the noise determines D. Now if the noise is thermal, we know that the ensemble will ultimately settle down to a Boltzmann distribution, $\sigma \propto e^{-R^2/2w}$, with w chosen so that the mean potential energy is $\frac{1}{2}k_\mathrm{B}T$. Since according to (13.11) the mean potential energy is $\frac{1}{2}\hbar\omega_0\bar{\xi}^2$, i.e. $\frac{1}{2}\hbar\omega_0 w$, it follows that

$$\sigma \propto e^{-R^2\hbar\omega_0/2k_\mathrm{B}T}.$$

For the quantized vibrator, isolated from thermal noise, the final distribution is e^{-R^2}, so that the intrinsic quantum noise has the same effect as if the vibrator were classical and subjected to noise from a source at temperatures $\hbar\omega_0/2k_\mathrm{B}$. If the losses were due to radiation we might be tempted to ascribe this noise to zero-point motion of the cavity modes, each of which has mean energy $\hbar\omega_0/2$. There is, however, nothing obviously analogous to zero-point motion in the model lossy mechanism that has led

79

to this result, and one would be wise to resist the temptation to interpret zero-point energy in classical terms.

The existence of quantum noise reveals the limitations of the dielectric model which in almost all other respects works extremely well. The defect is much the same as we have noted before in attempts to extend classical arguments too far – the dielectric model treats two interacting systems (resonator and molecule) as if they were separate entities, each acted upon by a determinate force due to the other. Thus the molecules are imagined as responding to a well-defined resonator field, and likewise the resonator to well-defined polarization oscillations of the molecules. We may expect, by analogy with the earlier examples, that on treating the coupled systems in a consistently quantal way new statistical features will enter. This is indeed the case, with fluctuations of amplitude and phase appearing quite independently of any external cause.

25, 108

It should be remembered that the noise temperature derived here is valid only for a perfectly cold system, since the molecules responsible for dissipation are all unexcited. A lossy resonator with walls at temperature T_0 may be simulated by a stream of untuned molecules for which the flux of unexcited molecules is $\Phi_d\, e^{\hbar\omega_0/k_B T_0}/(e^{\hbar\omega_0/k_B T_0}-1)$ and of excited molecules $\Phi_d/(e^{\hbar\omega_0/k_B T_0}-1)$; there is a net excess Φ_d of unexcited molecules to give the required dissipation, and the ratio of the two fluxes is the Boltzmann factor for temperature T_0. All molecules contribute equally to the fluctuations, which are therefore increased in proportion to the total flux, $\Phi_d \coth(\hbar\omega_0/2k_B T_0)$, so that the equivalent noise temperature T_N takes the form:

$$k_B T_N = \tfrac{1}{2}\hbar\omega_0 \coth(\hbar\omega_0/2k_B T_0). \tag{62}$$

At low temperatures the residual noise is purely quantal, with $T_N = \hbar\omega_0/2k_B$, and as the temperature is raised it increases until $T_N = T_0$ and the noise can be considered to be entirely thermal and to have the magnitude expected from classical arguments (cf. Johnson noise). The form of (62) matches Planck's expression for the mean energy of a harmonic vibrator, with the zero-point energy added, and the transition from quantum noise at low temperatures to classical noise at high is precisely parallel to the transition from zero-point energy of the vibrator to the classical equipartition result. At any temperature the equilibrium form of the σ-distribution is Gaussian, with a mean energy of $k_B T_N$ which is just the result of Planck. It is satisfactory to find, albeit in a special case, confirmation of the view endorsed in chapter 15, that randomness is communicated, and statistical equilibrium established, in the interaction of quantized systems.

I.158

Finally we note that when the vibrator is acted upon by a determinate force the response of the equivalent classical ensemble is exactly the same as described in chapter 13, every member responding independently. We therefore can give a complete prescription for finding how $\mathcal{P}(\xi, t)$ develops for a quantized vibrator acted upon by a determinate force and subject to dissipation from perfectly cold absorbers: $\mathcal{P}(\xi, t)$ is the projection on the

ξ-axis of a two-dimensional distribution $\sigma(\boldsymbol{R}, t)$ which develops according to (61), spins about the origin at angular velocity ω_0, and in this spinning state is displaced by a force $F(t)$ which gives rise to a bodily shift, always in the κ-direction, at a rate $\dot{\kappa} = F/(m\hbar\omega_0)^{\frac{1}{2}}$. In laboratory co-ordinates, x and $\dot{x}/\omega_0$ instead of ξ and κ, (61) takes the form:

$$\dot{\sigma} = (\hbar/4mQ_r)\nabla^2\sigma + (\omega_0/2Q_r)\,\mathrm{div}\,(\sigma\boldsymbol{r}). \tag{63}$$

Any applied force $F(t)$ moves the spinning σ-distribution bodily in the y-direction at a speed $\dot{y} = F/m\omega_0$. The projection of the distribution on to the x-axis gives $\mathcal{P}(x, t)$.

It is worth noting that when Q_r is large and F is a force applied at the resonant frequency, the force may be resolved into two rotating components, each of magnitude $F/2$, one of which is synchronized with the rotating σ-distribution, while the counter-rotating component produces only a tiny perturbation and may be disregarded. In the spinning frame, therefore, the distribution suffers a steady drift in a constant direction at a velocity $\boldsymbol{F}/2m\omega_0$. This adds a term $(-\boldsymbol{F}/2m\omega_0)\cdot\mathrm{grad}\,\sigma$ to the right-hand side of (63), which is equivalent to replacing the second term in this equation by $(\omega_0/2Q_r)\,\mathrm{div}\,[\sigma(\boldsymbol{r}-\boldsymbol{r}_0)]$, where $\boldsymbol{r}_0 = Q_r\boldsymbol{F}/m\omega_0^2$, the classical value for the amplitude at the resonance peak. The steady state is then a Gaussian peak, of width appropriate to the temperature, displaced bodily from the origin; and if $k_BT_0 \ll \hbar\omega_0$ this is exactly the coherent state which is thus seen to be less artificial in conception than might have appeared from the way it was introduced.

Dissipation and fluctuations in the maser

The general ideas developed in the last section are immediately extendable to the spontaneously oscillating maser, in which two processes are at work independently, the regenerative action of excited, well-tuned, molecules and the dissipative action of untuned, unexcited molecules. The excited molecules cause ρ_{ln} to climb upwards, according to (41), while the unexcited oppose the climb, according to (49); the two processes acting together result in an equation for the development of the diagonal terms:

$$\mathrm{d}\rho_{nn}/\mathrm{d}t' = -C_n\rho_{nn} + C_{n-1}\rho_{n-1,n-1} - n\rho_{nn} + (n+1)\rho_{n+1,n+1}, \tag{64}$$

in which $t' = t/\tau_e$ as before; $C_n = \tau_e\Phi\mu_{nn}$ and represents the number of molecules that are de-excited in time τ_e. From (38),

$$C_n = \tau_e\Phi\gamma^2(n+1)\sin^2\Omega_nT/\Omega_n^2. \tag{65}$$

The rate of climb is determined by C_n, which in essence is the same as the negative loss $-\kappa_e''$ of the dielectric treatment. At low values of n, $C_n \mathbin{\widetilde{\propto}} n$ and the excited molecules respond linearly to the resonator field. The critical flux, at which maser action begins, is such that $C_n = n$ when n is small, so that the first and second pairs of terms in (64) have equal and opposite effects; hence $\Phi_c = 1/\tau_e\gamma^2T^2$ in agreement with (7). We shall not worry about the precise variation of C_n with n since this is influenced by the unknown spread of residence times. It is enough to recognize that the initial proportionality to n breaks down and eventually C_n reaches, or even goes through, a maximum. The steady state of oscillation lies around the value of n at which $C_n = n$.

Before discussing the oscillatory steady state, however, let us examine the behaviour before maser action begins, and immediately after F exceeds unity.† So long as linearity prevails, with $C_n = Fn$ and F constant, the procedure that led from (45) to (60) and (61) applies equally well to (64), and allows the development of $\sigma(\boldsymbol{R})$ to be expressed as a differential equation:

$$\frac{1+F}{4\tau_e}\nabla^2\sigma + \frac{1-F}{2\tau_e}\operatorname{div}(\sigma\boldsymbol{R}) = \partial\sigma/\partial t. \qquad (66)$$

The drift back to the origin is now governed by a time-constant $2\tau_e/(1-F)$ which exhibits clearly the opposition of the dissipative and the regenerative processes, balanced when $F = 1$. On the other hand the diffusive process involves the cooperation of both, as shown by the extra factor $(1+F)$. If $F < 1$ the steady state has the Gaussian form of the ground state of the oscillator, centred on the origin, but spread out by a factor $[(1+F)/(1-F)]^{\frac{1}{2}}$:

$$\sigma \propto e^{-R^2/2w}; \qquad w = \tfrac{1}{2}(1+F)/(1-F). \qquad (67)$$

When F is raised above unity, the distribution explodes:

$$\left.\begin{aligned} &\sigma \propto w^{-1}\, e^{-R^2/2w}, \\ \text{and} \qquad &w(t) = w_0\, e^{(F-1)t/\tau_e} - \tfrac{1}{2}(F+1)/(F-1). \end{aligned}\right\} \qquad (68)$$

So long as $C_n = Fn$ the exponential spread of the distribution continues, with the maximum of σ remaining at the origin. As n increases, however, and C_n rises less rapidly than n, the outer regions of σ begin to slow down, while the inner regions continue to spread. Ultimately, when $C_n \sim n$ at the outside, the leading edge stops and the inner regions pile up against it, until the steady state has σ in the form of a ring.

At this point the σ-representation shows a weakness. The derivation of (66) from (64) depends on the validity of the linear approximation; as soon as C_n ceases to be proportional to n it becomes impossible to eliminate n in deriving a differential equation for σ. This does not mean that (66) must be utterly discarded, but only that it must be treated with caution as at best approximate, resting on the hope that to allow the coefficients to vary with R will take care of most of the problems. We shall find that the

† It should be remembered that the model is unrealistic at this point, since in the real maser the rate of growth is limited by T rather than τ_e. But the picture of how the development proceeds is probably fairly reliable apart from the time scale and is, in any case, introduced primarily to exhibit certain limitations of the σ-representation.

steady-state solution is indeed consistent with slowly varying drift and diffusion coefficients. We shall, however, also find that different diffusion coefficients are needed for radial and tangential variations of the distribution.

To discuss the radial variation first, it is enough to note that in the steady state ρ_{ln} is diagonal and governed by (64). Let us define $\bar{n}$ as $\Sigma_n n\rho_{nn}$; then

$$\mathrm{d}\bar{n}/\mathrm{d}t' = \sum_n n\,\mathrm{d}\rho_{nn}/\mathrm{d}t' = \sum_n (C_n - n)\rho_{nn} = \bar{C}_n - \bar{n}. \tag{69}$$

On the assumption of a narrow distribution, $\bar{n}$ in the steady state $= n_0$, where $C_n(n_0) = n_0$. In the vicinity of n_0 let $C_n = n_0 - C'_n(n - n_0)$, C'_n being the (positive) value of $-\mathrm{d}C_n/\mathrm{d}n$ at n_0. Then (69) may be rewritten

$$\mathrm{d}\bar{n}/\mathrm{d}t' = (n_0 - \bar{n})(1 + C'_n), \tag{70}$$

showing that $\bar{n}$ approaches n_0 exponentially with time constant $\tau_e/(1+C'_n)$. In terms of Q_{sat},

$$Q_{sat}/Q_r = 1/(1 + C'_n). \tag{71}$$

The same Q_{sat} determines the spread of n in the steady state. If $\mathrm{d}\rho_{nn}/\mathrm{d}t' = 0$ and ρ_{nn} is expanded as a Taylor series about n_0, (64) takes the approximate form

$$(1 - C'_n)(\rho_{nn} + y\rho'_{nn}) + n_0\rho''_{nn} \doteqdot 0, \text{ where } y = n - n_0.$$

As usual, the solution is Gaussian, $\rho \propto \mathrm{e}^{-y/2w}$, and $w = n_0/(1 + C_n)$; the mean square width $= n_0 Q_{sat}/Q_r$.

Let us translate this into a distribution for σ, which will be sharply peaked round a ring of radius $(2n_0)^{\frac{1}{2}}$. Each term in ρ_{nn} is represented by an axially symmetrical distribution which would be a perfectly sharp ridge at $(2n)^{\frac{1}{2}}$ if $\psi_n^2(\xi)$ followed the classical probability pattern exactly. As we have seen, however, the ridge is not perfectly sharp, and inside it there is a series of subsidiary rings, alternatively positive and negative. Now when we add all the patterns due to the ψ_n^2 in a range of about $n_0^{\frac{1}{2}}$, the radii of the ridges extend over a range of about $n^{\frac{1}{6}}$ times the widest separation of the subsidiary rings in any one pattern, say 10 times if $n = 10^6$. This is enough to ensure, first, that the subsidiary oscillations cancel each other and, secondly, that the resulting ridge is wide enough for the width of an individual constituent to be unimportant. That is to say, the ring in σ reflects very closely the distribution ρ_{nn}. Since each n appears as a ridge of radius $R = (2n)^{\frac{1}{2}}$, a range Δn is translated into a range $\Delta R = \Delta n/(2n)^{\frac{1}{2}}$; and if $\overline{\Delta n^2} = n_0/(1 - C'_n)$, $\overline{\Delta R^2} = 1/2(1 + C'_n)$, which is the mean square width of the coherent state, $\frac{1}{2}$, increased by the factor Q_{sat}/Q_r.

This is a good approximation to the solution of (64) in the steady state. Let us see what drift and diffusion coefficients are needed to reproduce it, for which purpose we return to (64). If only the last two terms were present the vibration would decay exponentially, and the σ-distribution would relax towards the origin with local velocity $-\boldsymbol{R}/\tau_a$. With the amplifying terms added the local velocity is reduced by a factor $(n - C_n)/n$, which of course vanishes at $n = n_0$. In the vicinity of n_0, where $C_n = n_0 - C'_n(n - n_0)$, the drift velocity is $-R_0(1 + C'_n)(n - n_0)/n_0\tau_a$, or $-2(1 + C'_n)(R - R_0)/\tau_a$ since $n \propto R^2$. The flux is σ times this, and represents a drift of the distribution towards R_0 from above and below. It is counteracted by the diffusion due to random noise. If $\sigma \propto \exp[-(R - R_0)^2/2w]$, the diffusive flux $-D\, d\sigma/dR$ is $(R - R_0)D\sigma/w$, i.e. $2(1 + C'_n)(R - R_0)D\sigma$ for w equal to $1/2(1 + C'_n)$ as determined above. In the steady state the fluxes must balance and therefore D must take the value $1/\tau_a$, or $1/2\tau_e$, which is twice as great as in the free vibrator. If we care to attribute the noise to sudden impulses occurring every time a quantum is transferred, it is clear that in the steady state the number of quanta communicated to the resonator by the molecules must equal the number removed, i.e. so far as noise is concerned the effective value of F is always unity in the neighbourhood of the steady state. It would be unwise to view this interpretation as anything but a convenience; there is nothing in the quantum mechanics to justify the semi-classical concept of sudden switches of state giving rise to sharp impulses. Nevertheless it works, as does the classical modelling on which we have laid so much emphasis, and there is no objection to using these ideas in solving problems provided one does not fall into the habit of believing that they describe what could really be observed, if only our senses were delicate enough to follow the processes – this is the slippery slope leading to hidden-variable interpretations of quantum mechanics, which as yet have made no useful contribution to advancing the progress of physics.

To evaluate the phase-diffusion caused by quantum noise, it is necessary to translate the non-conserved off-diagonal elements of ρ_{ln}, as expressed for example by (43), into a diffusion coefficient. Consider a narrow ring of radius R in the σ-representation, round which σ varies in an arbitrary manner. In a Fourier analysis of this variation, the term in $e^{i\nu\theta}$ represents a linear phase variation of the form $e^{i\nu s/R}$, if s is measured along the perimeter of the ring. Now such a variation decays exponentially as a result of diffusion, the diffusion equation,

$$D\partial^2\sigma/\partial s^2 = \partial\sigma/\partial t,$$

being satisfied by $\sigma \propto \exp[i\nu s/R - D\nu^2 t/R^2]$, so that the time-constant is $R^2/D\nu^2$. A time-constant varying as $1/\nu^2$ is exactly what the real part of (43) predicts, and comparison of the two expressions shows that when the system is perfectly tuned, so that $\Omega_n^2 = \gamma^2(n + 1)$,

$$D = \Phi\gamma^2 T^2 R^2/8(n + 1) = \tfrac{1}{4}\Phi\gamma^2 T^2 = F/4\tau_e, \text{ from (7).} \tag{72}$$

This is the tangential diffusion coefficient due to the excited molecules alone; the unexcited molecules have already been treated by a slightly different method and have yielded in (57) a time-constant of $8\bar{n}\tau_e/\nu^2$, i.e. $4R^2\tau_e/\nu^2$, equivalent to $D = 1/4\tau_e$. The diffusive processes are independent, both making contributions of the same sign to the real part of (43), so that the resultant tangential diffusion coefficient is $(1 + F)/4\tau_e$, the same as for both radial and tangential diffusion in the initial growth as governed by (66). Nothing has been said at this point, however, about n being small and, in contrast to radial diffusion, the tangential diffusion coefficient does not fall off as the steady state is approached. Here is additional reason for treating the σ-representation with caution; it gives an excellent general picture of the true quantal behaviour but needs tinkering here and there if it is to be used quantitatively – and the tinkering, however minor, is of such a nature as to make one sceptical about deriving any deep meaning from the model.

Measurement of maser frequency

When a maser is connected to a counter to measure its frequency, the counter will respond every time the signal entering it passes through zero. The act of responding tells the observer that, for example, $\mathscr{E}=0$ at some instant which is rather well defined in relation to the period of oscillation. Possessing such information, the observer must set up a wave-packet $\sigma(\xi, \kappa)$, with a well-marked peak, to describe his knowledge of the maser. So long as the observer is content to let the maser and counter run unregarded, his ability to predict the phase at a later moment is limited by the spread of the σ-peak round its ring. Suppose that the R.M.S. angular width of the peak has increased to $2\Delta\theta$; he must be uncertain to this extent of the phase of the oscillation when at last he notes the response of the counter. The spread $\pm\Delta\theta$ implies that the counter may respond any time within the interval $\pm\Delta\theta/\omega_0$; hence the observer must be uncertain of the frequency to the extent of $\Delta\theta/\omega_0 t$. If $\Delta\omega_0$ is the R.M.S. uncertainty in the frequency, as measured by counting cycles for time t,

$$\Delta\omega_0/\omega_0 = \Delta\theta/\omega_0 t. \tag{73}$$

Now tangential diffusion leads to a Gaussian distribution $e^{-s^2/4Dt}$, with $\Delta s=(2Dt)^{\frac{1}{2}}$; therefore $\Delta\theta=(2Dt)^{\frac{1}{2}}/R=(Dt/\bar{n})^{\frac{1}{2}}$, and

$$\Delta\omega_0/\omega_0=(D/\bar{n}\omega_0^2 t)^{\frac{1}{2}}. \tag{74}$$

We shall discuss this result in the next section, but must first draw attention to a point which has been so far ignored, the imaginary term in (43) which vanishes in a perfectly tuned maser. If $\omega_r \neq 0$ every excited molecule induces a phase shift to the off-diagonal terms in ρ_{ln}, proportional to ν, and the stream of molecules gives rise to a rate of phase shift Φ times as great. This is equivalent to a change in the angular velocity of the σ-representation, which will produce ν times the change in the oscillation frequency of the νth Fourier component projected onto the ξ-axis. Self-consistency demands that ω, the resultant frequency in the steady state, is such that the shift from ω_0, which is $2(\omega_e-\omega_r)$, is correctly given by (43):

$$2(\omega_e-\omega_r)=(\Phi\gamma^2\omega_r/4\Omega_n^3)[2\Omega_n T-\sin(2\Omega_n T)],$$

so that
$$\omega_r/\omega_e=\{1+(\Phi\gamma^2/8\Omega_n^3)[2\Omega_n T-\sin(2\Omega_n T)]\}^{-1}.$$

Apart from a change of notation, this is the same as (11). The dielectric model of the molecular beam gives as satisfactory an account of the frequency control as it does of the amplifying process. Only in the question of fluctuations does it fail.

The maser as frequency standard

A frequency standard is in essence an oscillator whose frequency is as little subject to perturbation as possible, whether from internal or external causes. A pendulum clock is limited in constancy of time-keeping by variations of pendulum length (temperature, wear of support), of g if it is moved to a different location, and of amplitude. Electrical oscillators such as maintained quartz crystal vibrators similarly suffer from temperature changes and secular drifts of the components in the maintaining circuit. On top of these effects, which are minimized by good design and maintenance, there is the intrinsic limitation of phase drift resulting from internal noise. It was noted in chapter 11 that a feedback oscillator suffered random walk in its phase, θ; after time t, when (11.44) is rewritten terms of $\Delta\theta$, the R.M.S. deviation,

$$\Delta\theta = (t/2\pi I_\omega)^{\frac{1}{2}}/\omega_0 L q_0,$$

where I_ω is the spectral intensity of voltage fluctuations in the circuit. If these fluctuations are due to Johnson noise in the resistor, at temperature T_0, $I_\omega = 2Rk_B T_0/\pi$ from (6.49), and **I.92**

$$\Delta\theta = (k_B T_0 t/E\tau_e)^{\frac{1}{2}} \quad \text{or} \quad \Delta\omega_0/\omega_0 = (k_B T_0/Pt)^{\frac{1}{2}}/Q_r, \tag{75}$$

in which E is the energy of oscillation, $q_0^2/2C$, P is the power dissipation, E/τ_e, and $\tau_e = L/R$.

To compare this with the maser we must first examine the relevance of (72) as an expression for the diffusion coefficient. This only enhances the dissipative contribution, as in (61), by a factor $F+1$, so that the equivalent noise temperature is $\frac{1}{2}(F+1)\hbar\omega_0/k_B$, still much less than room temperature unless F is very large indeed. In saying this it is taken for granted, in the light of the earlier discussion, that the buffering action of the molecules operates with quantum as with thermal noise. The conclusion is that thermal noise is the only source worth considering and that D, instead of being $1/4\tau_e$, would be $(2k_B T_0/\hbar\omega_0)/4\tau_e$, i.e. $k_B T_0/2\hbar Q_r$, in the absence of buffering. But since every step in the random walk due to noise is in practice reduced by a factor τ_e/T, and since D is proportional to the square of the step length, we have that

$$D = k_B T_0 Q_r/2\hbar Q_m^2, \tag{76}$$

and from (74),

$$\Delta\omega_0/\omega_0 \sim (k_B T_0 Q_r/2Q_m^2 E\omega_0 t)^{\frac{1}{2}} \sim (k_B T_0/2Pt)^{\frac{1}{2}}/Q_m. \tag{77}$$

The approximate signs are intended to show that no great care has been taken in relating T to Q_m, so that a numerical factor of the order of unity may well appear in a more meticulous treatment.

In the ammonia maser the residence time is about 10^{-4} s and Q_m about 10^7. A flux of 10^{13} molecules per second carries about 10^{-10} W, which is an upper limit for P; then if $T_0 = 300$ K,

$$\Delta\omega_0/\omega_0 \sim 5\times10^{-13}/t^{\frac{1}{2}} \ (t \text{ in seconds}).$$

In ten seconds the maser executes 2.4×10^{11} cycles and the uncertainty in frequency is about 2 parts in 10^{13}, equivalent to a phase uncertainty of 1/20 cycle, say 20°. This is very good indeed, but the hydrogen maser is still better.

The hydrogen maser[(5)] makes use of the transition in atomic hydrogen in which the electron and proton spins change their coupling. Both particles have spin $\frac{1}{2}$ and couple to give either a triplet state with parallel spins or a singlet with antiparallel, the former being higher in energy. The energy difference corresponds to a frequency of 1.42 GHz – the celebrated 21 cm line which has played a key role in galactic studies by radio astronomy. In a magnetic field, as fig. 4(a) shows, the triplet is split into Zeeman components; but it does not need a very strong field to break the coupling – the

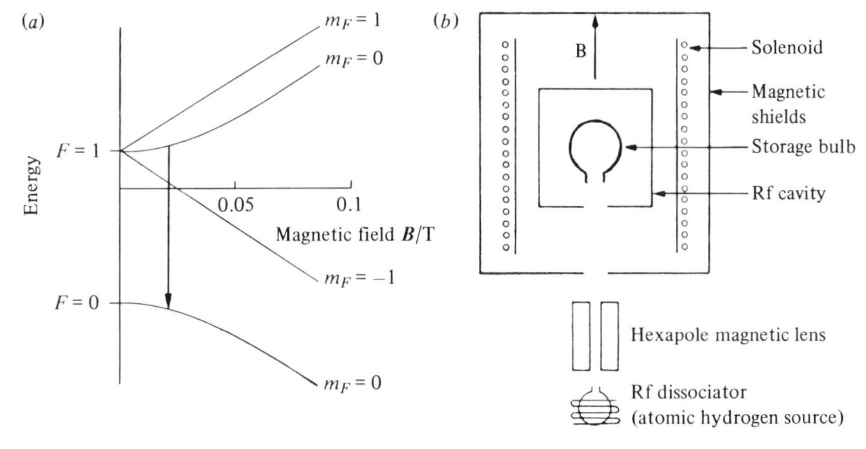

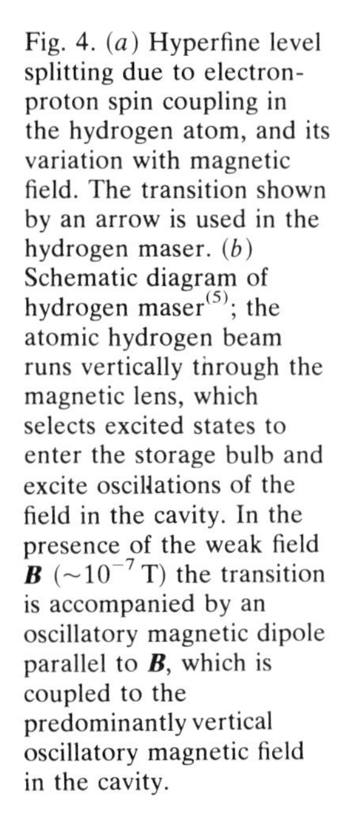
Fig. 4. (a) Hyperfine level splitting due to electron-proton spin coupling in the hydrogen atom, and its variation with magnetic field. The transition shown by an arrow is used in the hydrogen maser. (b) Schematic diagram of hydrogen maser[(5)]; the atomic hydrogen beam runs vertically through the magnetic lens, which selects excited states to enter the storage bulb and excite oscillations of the field in the cavity. In the presence of the weak field $\boldsymbol{B}$ ($\sim10^{-7}$ T) the transition is accompanied by an oscillatory magnetic dipole parallel to $\boldsymbol{B}$, which is coupled to the predominantly vertical oscillatory magnetic field in the cavity.

process is nearly complete at the right-hand side of the diagram. The details are unimportant for our discussion; what matters is that the upper two states ($F = 1$, $M_F = 0$ and 1), whose energy increases with the field strength, can be focussed by a magnetic lens (see fig. 4(b)) analogous to the electrostatic lens in the ammonia maser, while the lower two states are defocussed. In this way the transition shown can be used to stimulate oscillations of a resonant cavity by injecting a population that is largely in the excited state. What is special about the device is that the atomic beam does not pass straight through the cavity but into a silica bulb that is lined with teflon (polytetrafluoroethylene), in which the atoms bounce around until they manage to escape and are picked up by the vacuum pumps. Ramsey's intuition that such an arrangement could be made to work was brilliantly verified beyond all reasonable expectation – the coupling of the spins is so little perturbed by collision with the walls or with each other that the interaction between atoms and resonator field remains phase-coherent over very nearly one second, during which time an atom hits the walls several thousand times. Consequently Q_m may be as high as 5×10^9, perhaps 10^5 times the cavity Q. The obvious consequence is that temperature changes and other direct perturbations of the cavity affect the resonant frequency only 10^{-5} times as much as if the freely running cavity were used as a primary frequency standard. In this respect the maser has a great advantage over the quartz clock, but of course the atom or molecule which activates the maser must still be protected with the utmost care from disturbances to its environment. The very high Q_m also leads to high stability against noise; for a recent portable model[(6)] the theoretical estimate of $\Delta\omega_0/\omega_0$ is $4\times10^{-14}/t^{\frac{1}{2}}$ when t is measured in seconds, and this stability is realized over times ranging from 10 s to an hour. Extreme attention to detail in the design is needed to achieve this fine performance.

To appreciate the meaning of such a result it is helpful to consider a specific example. In an earlier test of the hydrogen maser principle, Vessot and Peters[(7)] constructed two masers which were tuned to a frequency difference of about 1 Hz. This is possible by applying slightly different small magnetic fields to the two bulbs, since at low fields the line frequency depends quadratically on field. The two outputs were mixed and the difference frequency amplified and fed to a timer which recorded the time (about 10 s) for ten cycles. A sequence of 19 measurements, occupying about 4 minutes, varied among themselves by only ±3 ms, less than 1/300 of the beat period. This means that over ten seconds the phase drift of each maser was somewhat less than one degree, of $\Delta\omega_0/\omega_0 \sim 10^{-13}$. The later model, mentioned above, is about 6 times better even than this.

21 The family of masers; from laser to travelling-wave oscillator

The ammonia and hydrogen masers are especially simple examples of maintained oscillators depending on the establishment of an *inverted population*. The general idea behind this expression can be understood by reference to an assembly of quantum systems in thermal equilibrium at some temperature T_0. The chance of finding any one system in a state of energy ε is given by the Boltzmann factor, $e^{-\varepsilon/k_B T_0}$, which is a monotonically decreasing function of ε. The radiation field due to the walls of the container, at T_0, passes through and interacts with the systems, but does not on the average disturb their equilibrium distribution between the energy levels. The upward transitions, stimulated by the radiation, from the more populous systems in lower-lying states are matched by the downward transitions from the less populous systems in higher states, which are enhanced in their activity since they emit spontaneously as well as by stimulation. If the distribution is modified by increasing the population of the higher levels at the expense of the lower, the coherent stimulated emission amplifies the radiation field more than stimulated absorption attenuates it; and the dominance of amplification is of course complete in the ammonia maser if only excited molecules are allowed to enter. This is an extreme example of an inverted population behaving as an active medium, that is, one that can amplify under appropriate conditions – in this case when the medium occupies a cavity resonator that is very closely tuned to the transition frequency of the molecules.

We have seen that the cavity plays only a minor role in controlling the frequency, the sharp molecular levels being far more effective for the purpose. If the active medium occupies a large volume, many wavelengths in extent, the problem of other dissipative mechanisms may vanish and any wave in the frequency bandwidth of the transition will be amplified as it passes through, by stimulated emission from the excited molecules. This is the most likely explanation of certain strong microwave emissions observed by radio astronomers, with brightness temperatures of the sources far exceeding that of the emitting gas cloud.[1] Such an excited gas cloud is intrinsically unstable, any spontaneously generated wave of the right frequency being amplified as it spreads out. Once the process has started, the amplitude soon rises above the ambient noise level, and the wave then provides the dominant mechanism for de-excitation, since the rate of de-excitation is proportional to the power level of the wave. By the same token any relatively strong signal entering the cloud from outside will assume the dominant role as it passes through. It is possible that the excited clouds are the product of a stellar explosion and that the expanding gas shell is not only maintained in an excited state by optical and ultra-violet radiation from the stellar remnant, but microwave radiation from the remnant may also be the primary source of the strongly amplified signal that ultimately emerges. The excitation will be effected by the higher energy photons, but only narrow bands of low energy radiation will be sufficiently well in tune with the relevant energy levels in the gas to be amplified.

There is no long-term coherence of the radiation produced in this way, since any primary source whose frequency lies in the bandwidth of the transition may be amplified; one expects to see an incoherent spectral line, possibly shifted and broadened by the Doppler effect in the expanding and turbulent gas cloud. If, however, the amplified wave after one passage through the cloud is fed back into it, it becomes the dominant source and will grow at the expense of other accidental sources, so long as the medium is replenished by whatever process is responsible for its excited state. Continuous cycling of the wave leads to coherence, though not necessarily to spectral purity. For a very strong wave may exhaust the medium so rapidly that only a short stretch behind the leading edge is amplified; in successive passages, then, the disturbance evolves into a short but intense pulse. This is a commonplace in high-power lasers[(2)] operating in the visible or near-visible spectral region, when Fabry–Perot interferometers or other optical resonators are used to cycle the radiation again and again through the active medium. For laser fusion machines fluxes of extraordinary intensity have been generated – 500 J in a pulse 500 ps long (15 cm in space), giving a flux of 10^{19} Wm^{-2} when focussed on to a target. This implies field strengths of 6×10^{10} Vm^{-1} and 200 T, and a radiation pressure not far short of 10^6 Atmospheres.†

In an optical laser the Fabry-Perot interferometer, or some other optical arrangement, replaces the resonant cavity of the maser. New problems arise from the very large number of resonant modes now available, some of them very closely spaced in frequency, so that the active medium may be able to support many. Even in low-power continuous lasers, where there is no sharpening of pulses by exhaustion of the medium, spectral purity is not automatic. This is but one aspect of laser design that will not be touched on here, and indeed the reader whose primary interest is in lasers should not seek enlightenment from the following pages, but from the multitude of specialist texts now available.[(3)] In the last chapter we examined the ammonia maser in some detail for its intrinsic interest as a soluble model of an important physical principle. At the end of the present chapter we shall look rather closely at a number of microwave devices and related

† It should be pointed out that in attempts at laser fusion by compression of pellets it is not the radiation pressure that does the work, but the still higher recoil pressure of the surface layers evaporated by the pulse.

processes whose analysis gives further substance to the argument, concerning the equivalence of classical and quantal treatments of the maser principle, which was emphasized in chapter 20. The following few pages are prescribed as an antidote to the sense of oppression which these slightly pedantic concerns may induce.

Mechanisms of population inversion

The direct separation by lens of excited from unexcited molecules is of very limited applicability. In this section we shall note some of the procedures for establishing an inverted population, with special reference to those that can be used for optical lasers. A large number of methods have been studied, with many specific examples of each put into practice, and an exhaustive catalogue will not be attempted. Instead we shall exhibit a few of the most important methods.

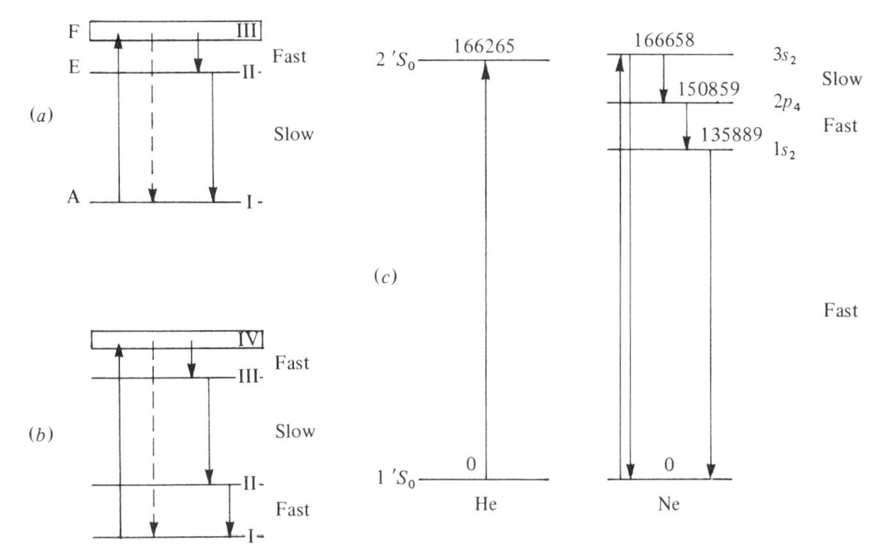

Fig. 1. (*a*) Schematic diagram of energy levels in a 3-level laser (ruby). (*b*) Energy levels for a 4-level laser (Nd in YAG). (*c*) Energy levels for the continuous He-Ne gas laser.

(1) *Optical pumping.* The active species is raised from its ground state, (I) in fig. 1(*a*), to an excited state (III) by irradiation at a higher quantum energy than will result from the laser action. The excited state must have a preference for decaying into a less highly excited state (II) from which spontaneous downward transitions are rare. The state (II) will then build up a substantial population; with strong enough pumping the ground state may be so depleted that its population falls below that of II and laser action becomes a possibility. Clearly a number of conditions must be satisfied, which may be summed up in the requirement that the active medium shall have a high fluorescence yield – the preferred route for de-excitation must involve an intermediate energy level. In ruby, the Cr^{3+} ions present as impurities have this property. The crystal absorbs green light to excite the

Cr^{3+} into the rather broad level labelled F in fig. 1(*a*); this is, however, not a very strong absorption and return to the ground state by re-radiation is much less likely than the emission of a number of phonons (quanta of lattice vibration) to attain the rather long-lived state labelled E. Intense pumping by a flash bulb can largely deplete the ground state and transfer a sizeable fraction of the ions to E in a time short compared with its lifetime of 3 ms. The mechanism for the fast radiationless† transition from F to E is somewhat obscure, but measurements of the number of red quanta (transitions from E to A) emitted for a given number of green quanta (A to F) absorbed leave no doubt about this route being highly preferred.

In contrast to the 3-level laser, severe depopulation of the ground state is not required for the 4-level laser (fig. 1(*b*)), of which Nd^{3+} in yttrium aluminium garnet or (more cheaply) a suitable silicate glass provide examples. The fluorescent level (III) here decays not to the ground state but to an excited state (II) which itself decays very rapidly, and whose energy is sufficient to preclude thermal excitation. In this way the population of ions in the lower level (II) involved in the laser process is kept very small.

(2) *Gas discharge.* Something of the same effect as with optical pumping may be achieved by passing a heavy current through an ionized gas. Brute force may produce results that could hardly have been predicted; as the highly excited molecules fall back to their ground state certain routes may turn out to be favoured and to involve, as in the four-level laser, a long-lived level followed by a short-lived. Inversion of the population is an automatic consequence. The helium-neon gas laser is more purposefully organized, its mode of operation being that shown schematically in fig. 1(*c*). An electrical discharge through a mixture of about 7 parts helium to one of neon excites helium atoms by electron collision to the state 2^1S_0 from which radiative transitions to the ground state are absolutely forbidden. It happens, however, that an excited state of neon, $3s_2$, has almost exactly the same energy relative to its own ground state, so that helium-neon collisions

† The transition from F to E is described as radiationless because the energy is emitted as lattice vibrations (phonons) rather than as a photon. Since the separation of the levels is many times the energy of any phonon that the lattice can support, a number of phonons must be emitted simultaneously. The probable mechanism may be visualized in quasi-classical terms by imagining it to have started, so that a Cr^{3+} ion is in a mixture of F and E states, with its charge distribution and the local surroundings vibrating rapidly. The frequency of vibration is, we know, much too high to generate an acoustic wave and so radiate a single phonon. But the crystal lattice is anharmonic. In particular, expansion and compression are not equivalent, and a process analogous to rectification takes place, the same process as is responsible for thermal expansion. In this case the vibrational motion around the ion has the effect of increasing the local pressure and initiating a shock wave. It is the shock wave that very quickly carries the energy away, and as there is no necessary limit to the intensity of shock, so there is no limit to the number of low frequency phonons that it can ultimately decompose into; the time taken for this last process is of no importance to the ion.

have a high cross-section for yielding this state of neon by exchange of excitation energy. From then on the neon behaves like a 4-level laser, making a transition fairly readily (10^{-7} s) to a lower level that decays considerably faster (2×10^{-8} s). It is the $3s_2$–$2p_4$ population that is inverted by this process. It should be noted that the $3s_2$ state is perfectly able to fall back to the ground state as well as to $2p_4$, but that trapping of the resonance radiation prevents depopulation of the excited state by this means. In this process the radiation emitted is reabsorbed by an unexcited neon atom which is thus raised to $3s_2$ to replace the originally excited atom. If there were no Doppler shift, the cross-section for resonant reabsorption would be $\frac{1}{2}\pi/k^2$, k being the wavenumber of the radiation; and

I.203 with n neon atoms per unit volume the mean free path of the photon would be $2k^2/\pi n$. Typically, at a pressure of 0.1 torr the mean free path would be 1 μm, but the Doppler shift increases this in the ratio of Doppler to natural line width, since only those atoms whose Doppler shift does not take them out of resonance will be effective absorbers. Even so, a photon is likely to suffer many processes of absorption and re-emission before escaping from the discharge, and during this time the atom that is temporarily excited may promote laser action.

(3) *Chemical excitation.* When hydrogen and fluorine combine to form hydrogen fluoride the first state of the molecule is one of high excitation. Left to itself the molecule loses energy by collision with others and the resulting rise of temperature is interpreted as the heat of reaction. But if the de-excited molecules are swept away the steady production of excited molecules may generate and maintain an inverted population. A laser can be operated in a steady state condition with continuous gas flow, H_2 and F_2 being admitted to the mixing chamber and HF extracted. The chemical process proceeds by the collision of fluorine atoms with hydrogen molecules, or hydrogen atoms with fluorine molecules, the two reactions forming a self-perpetuating sequence:

$$F+H_2 \rightarrow HF^* + H,$$

followed by

$$H+F_2 \rightarrow HF^* + F.$$

The stars indicate excited states. A little nitric oxide added to the hydrogen ensures the production of fluorine atoms when the streams mix: $NO+F_2 \rightarrow NOF+F$. The interaction between, for example, F and H_2 does not need anything like a head-on collision, and begins when the two are some distance apart, one of the hydrogens being attracted to the fluorine atom; thus HF and H may separate with an interatomic spacing in the HF considerably larger than the equilibrium separation. The molecule is left in a high vibrational state, and decays with a strong oscillatory dipole moment at the vibrational frequency which lies in the infra-red. It is this frequency which is coherently stimulated in the laser. For all its simplicity, however, the extreme chemical reactivity of the components makes this system unattractive.

No problem of removal of the de-excited molecule arises when excimers are used. There is no stable compound of krypton and fluorine, but KrF forms a bound molecule (excimer) in certain excited states. When an electrical discharge is passed through a mixture of krypton and fluorine, $Kr^* + F_2 \rightarrow KrF^*$, and the excimer decays very rapidly with the emission of a rather broad line in the ultra-violet, around 250 nm. The excimer can therefore be used to operate an ultra-violet laser which is tunable over the line-width. The excimer Xe_2 produces an even higher energy photon, at 170–176 nm; at this quantum energy the spontaneous de-excitation is so fast that the gas must be compressed to about 10 Atmospheres and very vigorously excited to create a sufficient population of excimers. The power dissipated by spontaneous emission is correspondingly high and demands excitation by a very intense (about 1 MW) electron beam. At such a level only pulsed operation is feasible, since most of the injected energy goes into heating the gas rather than into the coherent ultra-violet output. Nevertheless, very strong pulses of ultra-violet can be achieved.

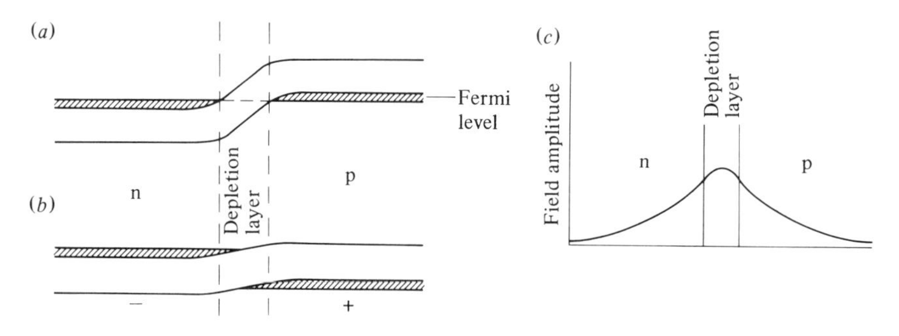

Fig. 2. (*a*) Impurity bands (shown shaded) of electrons in the conduction band of the n-type, and holes in the valence band of the p-type GaAs, separated by a depletion layer. In equilibrium the Fermi level is the same throughout the material. (*b*) A potential difference, as indicated by the + and − signs, allows electrons in the conduction band to fall into the valence band in the depletion layer. (*c*) Electrons and holes lower the dielectric constant except in the depletion layer. The amplitude profile shown speeds up the wave in the depletion layer and slows it down elsewhere, thus allowing a guided wave to travel with constant profile along the interface between n and p material.

(4) *The semiconductor laser.* Fig. 2(*a*) shows schematically the energy level structure of a p–n junction in a semiconductor such as GaAs. On the left there is a high enough doping level of tellurium donors for them to form a band of filled states just below the top edge of the energy gap, while on the right zinc atoms (acceptors) have been diffused in to overwhelm the donors so that the resulting p-type material has a band of vacant states just below the bottom edge of the gap. Without any external bias there is an electric field in the depletion layer between n and p, the Fermi level being the same on both sides. Application of a forward biasing voltage to reduce this field leads to the situation in fig. 2(*b*), and there is now a narrow region in which electrons in the conduction band coexist with vacancies in the valence band. This provides the population inversion, and de-excitation can occur by electrons falling into the vacancies and emitting a photon of wavelength 840 nm, in the near infra-red. The transition is rapid ($\sim 10^{-9}$ s) because in GaAs, unlike many other semiconductors, there is no need for a phonon to be emitted along with the photon to conserve crystal momentum. The material is transparent to light of the wavelength emitted, with a refractive index that is slightly lower in the p and n regions than in the

narrow active lamina between them. This allows confinement of the radiation in the vicinity of the lamina, as shown in the diagram; a wave travelling unchanged in form at a velocity intermediate between that of a plane wave in the lamina and that in the exterior regions must have a convex amplitude profile in the lamina to make it go faster, and a concave profile outside to make it go slower. This is the mode that can be caused to resonate by reflection to and fro between polished ends of the material, and to be amplified by stimulated transitions of electrons downwards into vacant states.

(5) *Surf-riding and bunching.* By these terms we describe a class of inversion procedures which operate through the interaction of a travelling wave with a stream of particles moving slightly faster than the wave, and being caused to slow down and thereby to transmit some of their energy to the wave. The amplification of acoustic waves, especially in conducting piezo-electric crystals, by passing a current exemplifies this process, as does the travelling-wave tube that is used to generate microwaves of millimetre wavelength. Such devices are normally analysed in a strictly classical manner, but their operation may equally well be considered as resulting from population inversion and stimulated emission. They therefore provide examples to link classical and quantal treatments, and for this reason deserve the fuller treatment to which we now turn.

Classical perspective on maser processes

The discussion of stimulated emission in chapter 20 showed that it can be considered as giving rise to an active dielectric medium such as, in a capacitor, would reveal itself as a negative resistance. To this extent the maser has an obvious resemblance to a negative-resistance oscillator and the parallelism was developed further in the analysis of fluctuations of amplitude and phase. One might attempt to reverse the process of thought and apply quantum mechanics to the oscillatory circuit, seeking a mechanism that will coherently refresh the circuit by providing photons to replace those dissipated by lossy processes; but a difficulty arises. In a typical maser the inverted population is relatively long-lived in the absence of stimulated transitions, spontaneous transitions being rather rare; on the other hand, in a negative-resistance circuit (or tunnel diode) there is steady dissipation at zero frequency and the stimulated transitions are commonly a rather minor perturbation of this continuing attempt to achieve equilibrium. Consequently, although it would probably not be out of the question to cast the theory of the oscillator into a form resembling that of a maser, it would amount to little more than an intellectual curiosity.

Electron-cyclotron oscillator

The same criticism does not apply to a class of oscillators exemplified by the klystron, the magnetron and the travelling-wave tube,[4] which can be understood in general terms from classical and quantal points of view with equal ease. For the purpose of illustration we shall examine the excitation of a microwave cavity by electrons executing cyclotron motion in a powerful magnetic field. Oscillators employing this mechanism have been realized,[5] as laboratory models rather than marketable devices, but we shall be little concerned with technical details, only with physical principles. The essential features are modelled by an electron cloud in two dimensions; each particle moves independently in a circular orbit in a plane normal to the magnetic field, $\boldsymbol{B}$, and is acted upon by a uniform electric field $\mathscr{E}$ rotating in the same sense as the cyclotron motion, with frequency ω. In practice the electrons, continuously injected with only a slight spread of energy around a mean of something like 20 kV, drift along $\boldsymbol{B}$ in helical paths, spending long enough in the resonant cavity, under the influence of $\mathscr{E}$, to execute many orbits. For mathematical convenience, on a point of no great physical significance, we shall assume an exponential distribution of residence times, with a probability $e^{-t/T}$ that an electron experiences $\mathscr{E}$ for a longer time than T.

I.237

Let us now calculate the impulse response function for the electrons distributed uniformly on the circular orbit of fig. 3 and consequently

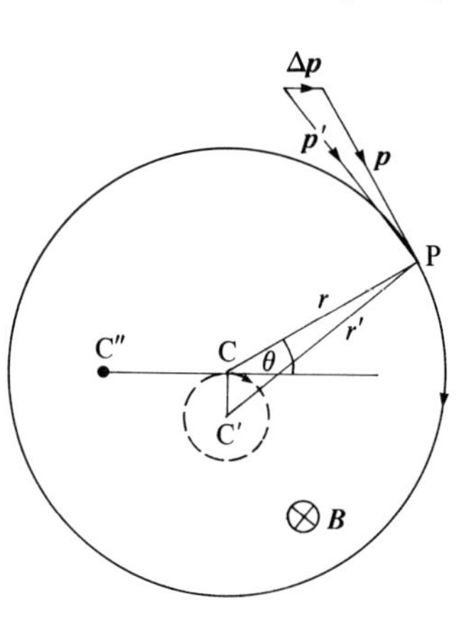

Fig. 3. Illustrating calculation of impulse response function for an assembly of electrons undergoing cyclotron motion.

exhibiting no resultant oscillatory dipole moment. The radius of the orbit, r, is p/eB for an electron having momentum $\boldsymbol{p}$, and this holds for relativistic as well as slow electrons. A typical electron, lying at P at the instant of the impulse (applied horizontally in the diagram), acquires momentum $\Delta\boldsymbol{p}$ from the impulse, so that $\boldsymbol{p}$ is changed to $\boldsymbol{p}'$ while the electron remains instantaneously at P. Its new orbit must be tangential to $\boldsymbol{p}'$ and have radius r' such that $r'/r = p'/p$; it follows by simple geometry that C′ lies vertically at a distance $\Delta p/eB$ below C, irrespective of the position of P. Hence the ring of electrons now begins to spin bodily about C, and the subsequent oscillations of dipole moment are as if the whole charge were initially concentrated at C and then set spinning at the cyclotron frequency in the orbit shown as a broken circle. In the absence of dissipation this impulse response leads to a loss-free contribution to the susceptibility, but decay as $e^{-t/T}$ introduces a lossy component, exactly as in chapter 8. In reaching this conclusion, however, we have assumed that all electrons have the same cyclotron frequency, $\omega_c = eB/m$, and have ignored the relativistic change of frequency with momentum. The change, small though it may be for these electrons of rather low energy, is nevertheless an essential feature of the device since it is only through its presence that the electrons may impart energy rather than absorb it.

When the impulse is applied, electrons near the top of the orbit in the diagram have their momentum increased, and their cyclotron frequency decreased, while those at the bottom have their momentum decreased and their frequency increased. If one were to observe the motion from a frame rotating at the mean frequency one would see electrons at the top and the bottom slowly drifting towards the left-hand side of the orbit, bunching together at the left and leaving a deficiency on the right. The centroid of the electrons would thus be seen to move to the left from C at an easily calculated rate. An electron at P has its momentum increased in magnitude by $\Delta p \sin\theta$, and its frequency by $\omega_c' \Delta p \sin\theta$, where $\omega_c^0 = \mathrm{d}\omega_c/\mathrm{d}p$, and is negative. This is its angular velocity as seen in the rotating frame, and the horizontal component of its velocity is $r\omega_c' \Delta p \sin^2\theta$, with a mean value for the whole ring of $\frac{1}{2} r\omega_c \Delta p$. After time t, therefore, the impulse response is represented not by the vector C′C, but by C′C″, where $\mathrm{CC}'' = \frac{1}{2} r\omega_c' t \Delta p$; at the same time, in the laboratory frame the vector C′C″ has rotated clockwise through $\omega_c t$. It should be noted that for a small impulse CC′ is correspondingly small, and that in time CC″ may well grow to a considerably greater magnitude without getting anywhere near the orbit radius, so that the linear approximation is not invalidated.

The position of the centroid after time t, expressed as a complex number, determines the impulse response function. As a result of an impulsive unit electric field, $\mathscr{E} = \delta(t)$, the momentum change $\Delta p = e$, so that $\mathrm{CC}' = 1/B$ and $\mathrm{CC}'' = \frac{1}{2} r\omega_c' e t = \omega_c' p t/2B$; if there are N electrons in the ring the response of the dipole moment takes the form:

$$h(t) = (Ne/B)(\mathrm{i} + \tfrac{1}{2}\omega_c' p t)\, \mathrm{e}^{-\mathrm{i}\omega_c t - t/T}; \qquad t > 0. \tag{1}$$

By (5.4), the susceptibility follows:

$$\chi(\omega) = \int_0^\infty h(t)\, \mathrm{e}^{\mathrm{i}\omega t}\, \mathrm{d}t = \frac{NeT}{B(1 - \mathrm{i}\delta T)}[\mathrm{i} + \tfrac{1}{2}\omega_c' p T/(1 - \mathrm{i}\delta T)], \tag{2}$$

in which $\delta = \omega - \omega_c$. Gain or loss is controlled by the imaginary part of χ,

$$\chi''(\omega) = \frac{NeT}{B(1 + \delta^2 T^2)}[1 + \omega_c' p \delta T^2/(1 + \delta^2 T^2)]. \tag{3}$$

The first, non-relativistic, term is always lossy, but the loss may be converted to gain by the relativistic bunching provided the second term is large enough and negative; δ must be positive, since ω_c' is negative, and the optimum tuning is such as to make $\delta T = \frac{1}{2}$. The residence time T must be long enough for $|\frac{1}{2}p\omega_c' T| > 1$. Now $p = mv/(1-v^2/c^2)^{\frac{1}{2}}$ and $\omega_c = \omega_{c0}(1-v^2/c^2)^{\frac{1}{2}}$, where m is the rest mass of the electron and $\omega_{c0} = eB/m$, from which it follows that $\omega_c' = -\omega_c v/E$, in which E is the total energy of the electron, including its rest energy mc^2. The criterion for amplification is that $\omega_c T > 2E/pv$. For electrons of only modest energy, $2E/pv$ is roughly the ratio of rest energy to kinetic energy, i.e. about 25 for electrons of 20 keV energy. Amplification should become possible, then, when $\omega_c T > 25$ and the residence time is long enough for more than four orbits to be accomplished; this presents no technical problem. When $\omega_c T$ is much greater than this critical value the tuning range in which amplification occurs runs from just above ω_c to $\omega_c + |\omega_c' p|$, though the factor outside the brackets in (3) limits the usefulness of the process when δ is large. It is clear, however, that elaborate precautions are not necessary to ensure uniformity of energy for the electrons, provided that they are used to excite a cavity tuned to a frequency slightly above the highest cyclotron frequency present, that of the least energetic electrons.

The quantal analysis of the same system starts conveniently with a single electron in its nth energy state, n being very large. We know that the behaviour closely corresponds to that of a harmonic oscillator, with the oscillator strength f equal to $n+1$ for upward and $-n$ for downward transitions to neighbouring levels, the only allowed processes. Because of the finite residence time the levels are broadened, with Lorentzian profiles if the distribution of lifetimes is exponential. When the electron is irradiated at such a frequency as will stimulate downward transitions, upward transitions will also take place unless the level spacings are different. In the non-relativistic case of equal spacings the larger f-number for upward transitions will in fact ensure that stimulated absorption outweighs stimulated emission, and the system will be lossy. It is only when relativistic effects cause the level spacing to decrease at higher energies that an irradiating frequency can be found, higher than ω_c, which will sufficiently favour the downward transitions that the net result is gain rather than loss. If the electrons had a Boltzmann energy distribution it would be necessary for the frequency to be higher than ω_c for the slowest, and the majority of processes would then involve just those electrons for which the relativistic advantage is least. By using a fairly homogeneous beam of slightly relativistic electrons all are enabled to contribute more or less equally to the gain.

To make the argument quantitative we note, from (13.39), that the impulse response function is expressed as the difference between two terms, but it must now be remembered that the frequency ω_0 is not quite the same in each term, being $\omega_c(n)$ for transitions between the levels $n+1$ and n. Thus we should write for a linear oscillator in a high quantum state, unlimited by finite resident time:

$$h(t) = \frac{\mathrm{d}}{\mathrm{d}n}[(ne^2/m\omega_c)\sin\omega_c t] = \frac{\mathrm{d}}{\mathrm{d}n}[(ne/B)\sin\omega_c t],$$

a factor e^2 being supplied so that $h(t)$ is the dipole response to unit impulsive electric field. The orbiting electron is represented by a two-dimensional oscillator, and the appropriate modification allows $h(t)$ to rotate and be expressed as a complex quantity:

$$h(t)=\frac{\mathrm{d}}{\mathrm{d}n}[(\mathrm{i}ne/B)\,\mathrm{e}^{-\mathrm{i}\omega_c t}]=(e/B)[\mathrm{i}+nt\,\mathrm{d}\omega_c/\mathrm{d}n]\,\mathrm{e}^{-\mathrm{i}\omega_c t}. \qquad (4)$$

We now multiply by $N\,\mathrm{e}^{-t/T}$ to allow for the number of electrons and their residence times; further we note that, since $n \propto p^2$, $n\,\mathrm{d}\omega_c/\mathrm{d}n=\frac{1}{2}p\,\mathrm{d}\omega_c/\mathrm{d}p=\frac{1}{2}p\omega_c'=\frac{1}{2}reB\omega_c'$. With these changes, (4) becomes identical to (1).

After all that has gone before it will not come as a surprise that the classical and quantal treatments agree so perfectly, since we are dealing in effect with a harmonic oscillator where this is to be expected. What is instructive is the demonstration that the process which preferentially encourages stimulated emission (in this case the unequal level spacing) is the same as induces bunching and permits the systematic development of a polarized state of the assembly, capable of energy transfer to the applied field. It is, of course, equally true that with a lower irradiation frequency, when δ in (3) is negative, the bunching process leads in both treatments to an enhanced absorption. Bunching is the classical expression of a mechanism of energy transfer that can be arranged to favour either absorption or emission. The conditions under which emission is favoured turn out to be such as to maximize the transfer rate at a point in a distribution where the population is inverted.

From the point of view of quantum mechanics there is little difference in principle between this system and a hydrogen atom which, under the stimulus of an oscillatory electric field, develops an oscillatory charge concentration by which energy transfer is effected. Unlike the hydrogen atom, however, the extended systems discussed here are equally amenable to classical analysis.

Bunching and maser action

A very similar bunching action can be seen as the classical mechanism by which acoustic waves in certain solids are absorbed, or in the right circumstances amplified, by interaction with conduction electrons;[6] and here too the process when viewed quantally is closely analogous to maser action. It is most marked with longitudinal compressional waves in piezo-electric semiconductors, such as cadmium sulphide, CdS, where the rarefactions and compressions give rise to an electric field pattern, parallel to the propagation direction and with sinusoidally varying amplitude. The wave

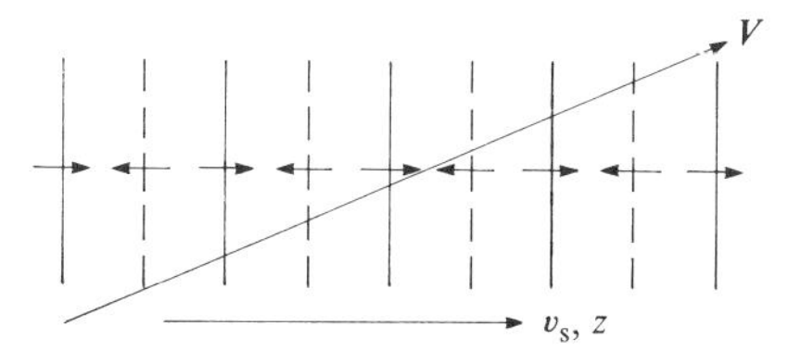

Fig. 4. Crests (full lines) and troughs (broken lines) of a compressional wave in a piezo-electric semiconductor and the resulting electric field pattern (short arrows); the path of an electron through this field pattern is also shown.

in the material lattice is thus accompanied by a wave of electrostatic potential, $V_0 \cos(qz-\omega t)$ if progagation is along the z-direction, as shown in fig. 4. Any electrons, excited thermally or optically, may interact with this potential and extract energy from, or transfer it to, the acoustic wave; but if the free path of the electrons between collisions is much longer than the acoustic wavelength, strong interaction is confined to those electrons whose velocity component v_z matches the sound velocity v_s, so that they move in a nearly constant field. This is obvious from a classical point of view, since any electron that travels through alternate crests and troughs is alternately accelerated and decelerated, with only small excursions of its energy from the mean. By contrast, for those electrons having v_z equal to v_s some are continuously accelerated and others continuously decelerated, and there is a real possibility of substantial energy transfer. The same condition arises naturally if one considers the wave as a quantized entity able to transfer energy in quanta of $\hbar\omega$ and momentum in quanta of $\hbar q$, provided these transfers are compatible with the conservation laws. When an electron receives a small increment, $\Delta \boldsymbol{p}$, of momentum its kinetic energy changes by $\boldsymbol{p}\cdot\Delta\boldsymbol{p}/m$, i.e. $\hbar q p_z/m$ if $\Delta\boldsymbol{p}=(0,0,\hbar q)$. Equating the energy change to $\hbar\omega$ yields the required result that p_z/m, which is v_z, must equal ω/q, which is v_s.

To illustrate the analogy between the classical bunching process and the maser principle it is enough to consider a simple one-dimensional model in which an electron in the state $\Psi_0=e^{i(kz-\Omega_k t)}$, where $\Omega_k=\hbar k^2/2m$, is perturbed by a potential, $V_0\cos(qz-\omega t)$, switched on at $t=0$. It is convenient to work in the frame in which the wave is at rest and the electron is moving at velocity $v=v_z-v_s$, having $k=mv/\hbar$. If $v>0$, as assumed in the calculation, the electron runs ahead of the wave in the laboratory frame, this being the condition for wave amplification. We concern ourselves with interactions so weak that the solution need contain only terms linear in V_0. The time-dependent Schrödinger equation:

$$(\hbar^2/2m)\partial^2\Psi/\partial z^2-eV_0\Psi\cos qz=-i\hbar\partial\Psi/\partial t, \qquad (5)$$

has in this approximation the solution

$$\Psi_1=\exp\{i(kz-\Omega_k t)\}+a_1(t)\exp\{i[(k+q)z-\Omega_{k+q}t]\}$$
$$+a_2(t)\exp\{i[(k-q)z-\Omega_{k-q}t]\}. \qquad (6)$$

Only at higher orders of V_0 are terms with wavenumber other than $k\pm q$ excited. When (6) is substituted in (5), and $\cos qz$ expressed as $\frac{1}{2}(e^{iqz}+e^{-iqz})$,

$$\dot{a}_1\exp\{i[(k+q)z-\Omega_{k+q}t]\}+\dot{a}_2\exp\{i[(k-q)z-\Omega_{k-q}t]\}$$
$$=(eV_0/2i\hbar)(e^{iqz}+e^{-iqz})\Psi_1$$
$$\sim(eV_0/2i\hbar)\{\exp\{i[(k+q)z-\Omega_k t]\}+\exp\{i[(k-q)z-\Omega_k t]\}\},$$

since the terms arising from $a_{1,2}$ in Ψ_1, are of second order in V_0. The terms in $(k+q)z$ and in $(k-q)z$ yield separate equations for $\dot{a}_1$ and $\dot{a}_2$, from which, since $a_{1,2}=0$ when $t=0$,

$$a_1=\alpha_1\{1-\exp[i\hbar(2kq+q^2)t/2m]\}$$
and
$$a_2=\alpha_2\{1-\exp[-i\hbar(2kq-q^2)t/2m]\}, \qquad (7)$$

where $\alpha_1 = meV_0/[\hbar^2(2kq+q^2)]$ and $\alpha_2 = -meV_0/[\hbar^2(2kq-q^2)]$. Hence from (6),

$$\Psi_1 = \Psi_0\{1+\alpha_1 e^{iqz}[e^{-i\hbar(2kq+q^2)t/2m}-1]+\alpha_2 e^{-iqz}[e^{i\hbar(2kq-q^2)t/2m}-1]\}. \tag{8}$$

If we take $\Psi_0^*\Psi_0$ as defining the initially uniform electron density, $\bar{\rho}$, its development follows from (8):

$$\rho(t)/\bar{\rho} = \Psi_1\Psi_1^* \sim 1+2(\alpha_1+\alpha_2)[\cos q(z-vt)\cos(\hbar q^2 t/2m)-\cos qz] + 2(\alpha_1-\alpha_2)\sin q(z-vt)\sin(\hbar q^2 t/2m). \tag{9}$$

This expression shows how the electron stream develops a periodic bunching which increases with time and drifts along, relative to the potential, at the mean electron velocity v. The classical limit, which is reached by letting either $\hbar$ or q/k tend to zero, follows from (9) on substituting for α_1 and α_2:

$$2(\alpha_1+\alpha_2) \sim -eV_0/mv^2$$

and

$$2(\alpha_1-\alpha_2) \sim 2meV_0/\hbar^2 kq.$$

Hence

$$[\rho(t)/\bar{\rho}]_{\text{Cl}} \sim 1-\frac{eV_0}{mv^2}[\cos q(z-vt)-\cos qz-qvt\sin q(z-vt)]. \tag{10}$$

This result is the same as is obtained by a classical calculation of the effect of the potential on a uniform stream of particles initially all moving with the same speed. The last term, with its coefficient proportional to t, grows without limit; it has its origin in the fact that when the potential is switched on, electrons at different points have the same kinetic energy but different potential energies, and therefore different total energies. As they migrate over the hills and valleys of potential their mean velocities are correspondingly different, so that in addition to the periodic ripple in ρ caused by electrons spending different times near crests and troughs of potential, there is a systematic growth in the bunching process as the faster electrons catch up the slower. This process, which will reappear in our discussion of the klystron, does not go on indefinitely, for as the faster electrons actually overtake the slower the bunches diffuse out again; but this is only apparent when higher orders of V_0 are included in the calculation. The smaller V_0 is, the longer the time before non-linear effects intervene.

Rearrangement of (10) shows more clearly how the peaks of ρ grow and shift relative to the potential:

$$[\rho(t)/\bar{\rho}]_{\text{Cl}} \sim 1-(eV_0/mv^2)(4\sin^2\tfrac{1}{2}X+X^2-2X\sin X)^{\frac{1}{2}}\sin(qz-X-\theta), \tag{11}$$

where $X = qvt$ and $\tan\theta = 2\sin^2\frac{1}{2}X/(X-\sin X)$. The amplitude and phase are shown in figs 5(a) and (b). Initially the bunching of ρ rises as t^2 and is strongest where $qz = \pi$, 3π etc., i.e. at the potential minima, as would

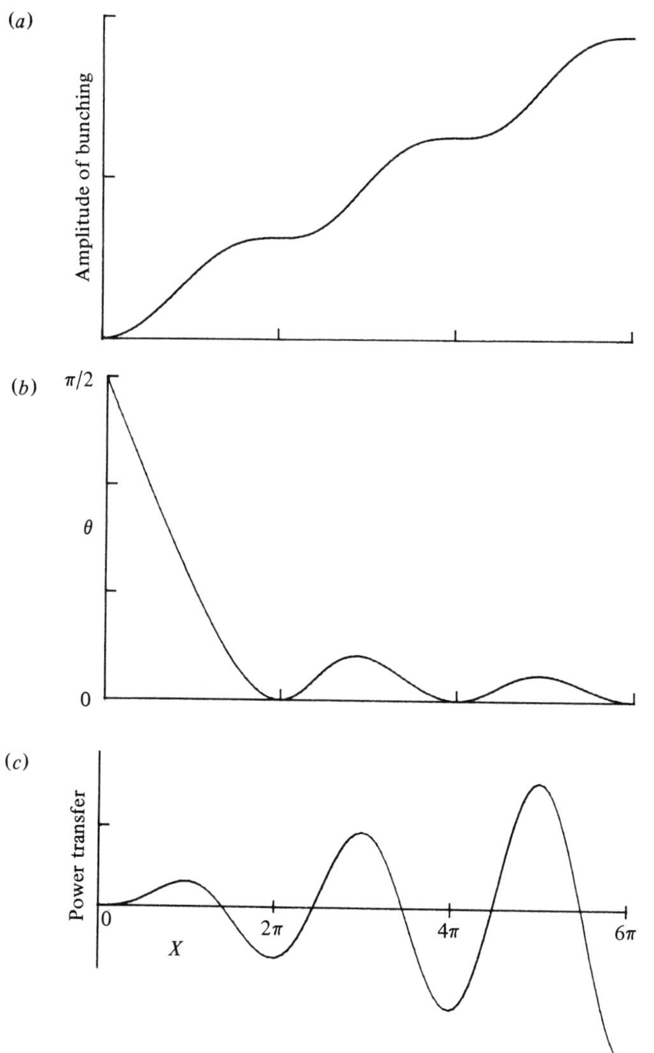

Fig. 5. Bunching of an initially uniform electron beam travelling slightly faster than a longitudinal wave field: (*a*) amplitude and (*b*) phase of density fluctuations at a distance X from the point of entry of the uniform beam; (*c*) rate of transfer of energy from electrons to wave.

be expected from elementary dynamical arguments. As time proceeds, the linear growth dominates and, as θ approximates to zero, the centres of the bunches move with those electrons that have the mean velocity v, having been at the outset at points of zero potential. Half these points act as foci of electron bunches, where those behind overtake and those ahead fall back; the other half conversely act as foci of rarefactions.

The rate of energy transfer, P, from the electrons to the wave is measured by the mean retarding force exerted by the fluctuating potential:

$$P \propto \langle -e\rho(\mathrm{d}V/\mathrm{d}z)\rangle \propto qe^2 V_0 \langle \rho \sin qz \rangle,$$

and is therefore determined by the magnitude and sign of the $\sin qz$

component of ρ. This follows immediately from (10):

$$P \propto \sin X - X \cos X, \tag{12}$$

and is shown in fig. 5(c). Obviously the oscillations of the curve reflect the migration of the bunches through alternate regions of retarding and accelerating force. In the initial stages the bunches, starting at potential minima, move forward to experience a retarding force, but the amplifying power of the electron beam is exhausted by the time $X = \pi$; if the electron velocity exceeds that of the wave by v, the useful time of interaction is not more than π/qv. This accords with the Uncertainty Principle, as the following argument shows, which holds in any frame of reference. For a real transition to occur there must be conservation of wavenumber and energy which, as we have seen, requires when $q \ll mv_z/\hbar$ that the electronic and acoustic velocities match. Now the wavenumber of the potential is well defined, but for an interaction time t there is an uncertainty in frequency, $\Delta\omega$, of roughly $1/t$; the perturbation experienced by the electron is therefore resolvable into Fourier components with velocities, ω/q, that spread over a range of $\Delta\omega/q$, roughly $1/qt$. Hence for a component to be present that satisfies the conservation rules, $1/qt$ must be at least comparable with v, the discrepancy in velocity, and the expected result follows: t must not be longer than something like $1/qv$. In oscillators that depend on an electron beam interacting with a wave the physical size of the device limits the interaction time, while with acoustic waves in solids it is the scattering of the electrons by defects that produces an equivalent effect, in this case more by rendering the electronic wavenumber slightly uncertain.

The argument has been presented for an electron travelling faster than the wave, so that the bunching runs ahead into a region of retarding force. Conversely, an electron moving more slowly than the wave lags behind and finds itself in a region of accelerating force; such an electron attenuates, rather than amplifies, the wave. Looked at from a quantum point of view the relative probabilities of exciting electronic states $k \pm q$ are governed by α_1^2 and α_2^2, i.e. $(2kq \pm q^2)^{-2}$ if $k = mv/\hbar$. When the relative velocity v is positive, k is positive and $\alpha_1^2 < \alpha_2^2$, the electron being more likely to lose than to gain energy; but the opposite is the case when $v < 0$. The difference between α_1 and α_2 becomes very marked when q, instead of being much less than k, is comparable. In particular, when $k = \frac{1}{2}q$, α_2 becomes infinite. An observer moving with the wave sees the electron suffer Bragg reflection. In the laboratory frame also the electron is reflected off the moving lattice formed by the wave, and it suffers an energy loss as a result of the Doppler shift of its frequency; correspondingly the wave benefits from the recoil. As we approach the classical limit the relative velocity v required for Bragg reflection goes to zero, and in any practical case $|k|$ may be taken as much greater than q. Then the probability of a downward transition exceeds that of an upward transition by $\alpha_2^2 - \alpha_1^2$, which now is proportional to $1/k^3$, i.e. $1/v^3$, becoming very strong as v approaches zero and changing rapidly from an excess of downward to an excess of upward excitations. If the wave is interacting with a stream of electrons of different velocities, it will be amplified only if the distribution of velocities provides more moving slightly faster than are moving slightly slower – in that range of distribution where $v_z \sim v_s$, the population must vary in the opposite way to a thermally equilibrated population, with more at higher energy than at lower. From the quantum viewpoint this is to ensure more downward than upward transitions; from the classical to ensure that the bunching is preferentially disposed towards retarding regions of the wave.

Having established once more the correspondence between bunching and maser action, let us note a few examples of waves interacting with electron streams, starting with the example that introduced the discussion.

Acoustic attenuation and amplification in solids[6]

In cadmium sulphide the electrons that are to interact with acoustic waves can be generated by the internal photoelectric effect, bound electrons in the valence band being excited optically into the conduction band where they soon become thermalized, with a Maxwell–Boltzmann distribution of velocities. The R.M.S. velocity at room temperature is something like 10^5 ms^{-1}, perhaps 50 times the acoustic velocity. Let us make the (unfortunately unrealistic) assumption that the mean free path of the electrons is very long; then the only electrons capable of exchanging energy with a given acoustic wave are those moving so as to keep pace with the wave. A wave moving from left to right in fig. 6 can exchange energy with electrons whose $\boldsymbol{k}$-vector lies near the interaction surface, a plane normal to the wave-velocity and displaced $mv_s/\hbar$ from the origin. Since the normal distribution of $\boldsymbol{k}$-values, as in fig. 6(a), has its greatest density at the origin, falling off steadily as k increases, there is a greater density to the left of the surface than to the right, and hence more processes involve absorption

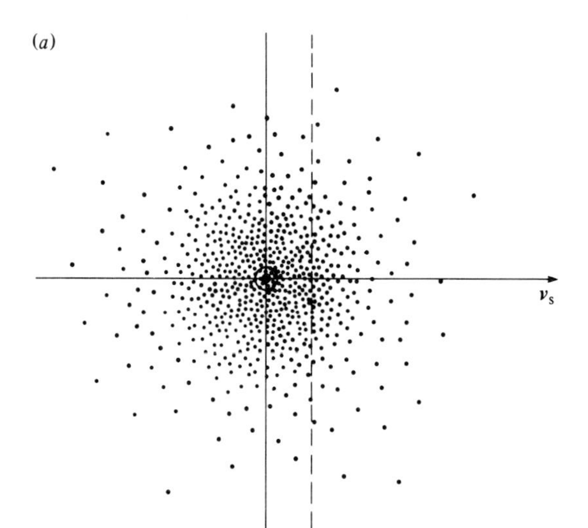

Fig. 6. (a) The dots represent a Maxwell–Boltzmann distribution of velocities (or $\boldsymbol{k}$) among the electrons; the broken line is the interaction surface on which the electrons stay on a wave-front of the acoustic wave. (b) An electric field has shifted the $\boldsymbol{k}$-distribution so that its centroid lies to the right of the interaction surface. (c) Attenuation in CdS as a function of longitudinal electric field strength[7]. At the left, where $\mathscr{E} > 700$ Vcm^{-1}, the wave is amplified.

Fig. 6 (*cont.*)

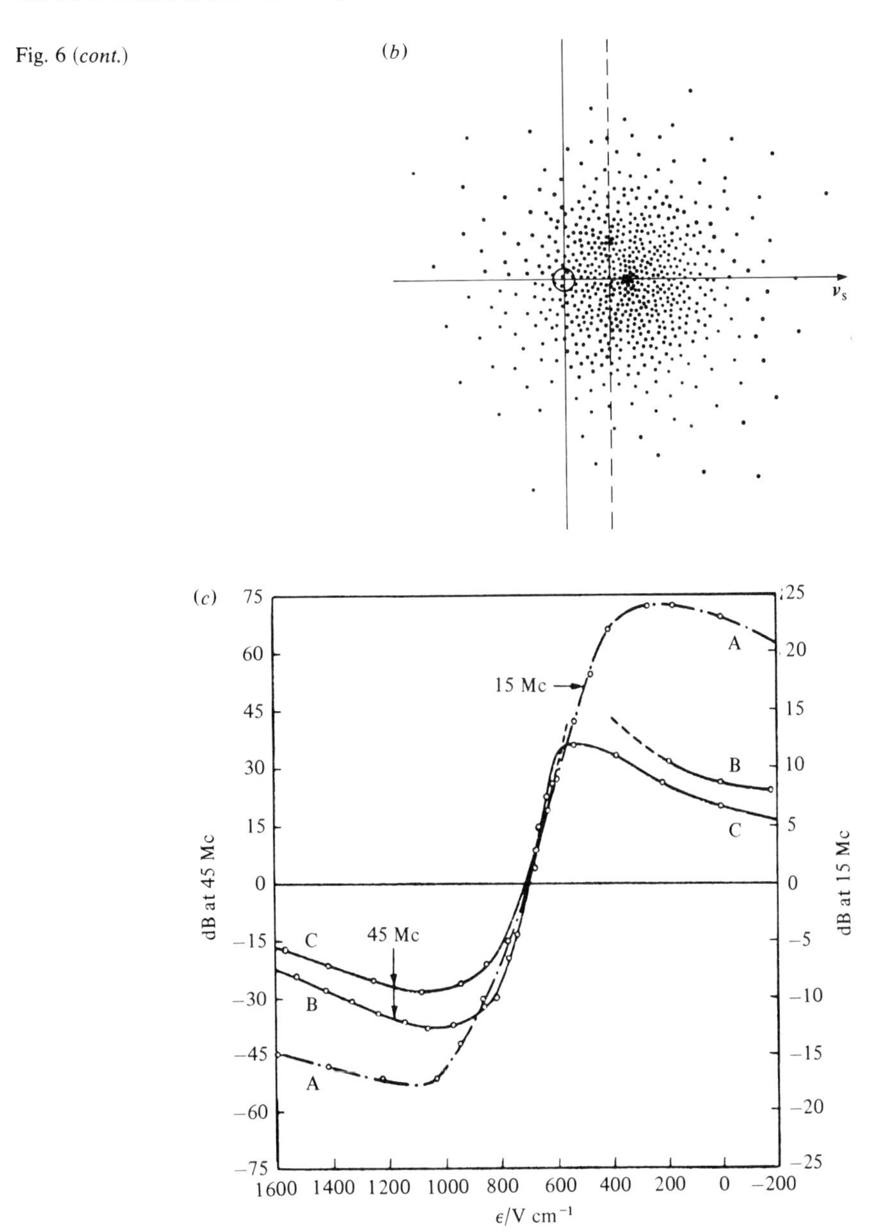

of energy from the wave than emission to it; this accounts for the acoustic attenuation. By applying a strong electric field, however, it is possible to shift the distribution very considerably, as indicated in fig. 6(*b*). This has been drawn to show a bodily shift, but it is likely that the distribution will be broadened asymmetrically at the same time. Leaving such a comparatively minor detail aside, it is clearly possible in principle to arrange that the density is greater on the right than on the left, and thus to convert acoustic attenuation into amplification by an excess of emission processes over absorption.

This analysis is too primitive to explain the details of the observation of the predicted effect shown in fig. 6(*c*), since the requirement of a long mean free path is very far from attainment, and interaction is not limited to electrons lying near the interaction surface. The shape of the curve can be derived rather well by not very involved classical arguments based on solution of the conduction equation for fields varying in space and time;[8] by contrast, a genuinely quantal derivation, incorporating frequent scattering processes into the wave equation, would be formidably difficult. Not only that, it might well hide the fundamental point that the transition from attenuation to amplification when the drift velocity exceeds the velocity of sound is an almost inevitable consequence of any theory in which the wave and the electron assembly are the only important constituents. So long as the steady electric field serves mainly to impart a drift velocity to the electrons, without grossly modifying their distribution as observed in a frame moving with the drift velocity, an observer in this frame will see the wave attenuated by interaction with the electrons; to this observer, attenuation is the only process that can increase the entropy and therefore the only possible process. Now in the laboratory frame the wave and the electrons may be moving from left to right, but if the electron drift exceeds v_s the moving observer sees the wave proceeding to the left, and will find that its amplitude is less on the left than on the right. This implies that in the laboratory frame it is amplified as it progresses.

The experimental result illustrated in fig. 6(*c*) shows the device as an amplifier, but it is worth noting that it can also oscillate spontaneously. Because of the form of the curve an acoustic wave travelling with the electron stream is amplified more strongly than one travelling in the reverse direction is attenuated, and a standing wave therefore may suffer net amplification. Although coherent oscillation was not observed, it was found that in the absence of an acoustic input bursts of noise built up, the time scale making it clear that many to-and-fro traverses were required before the amplitude finally reached saturation, with the extraneous dissipation matching the input from the electrons. Unlike the ammonia maser and others, the energy source has a continuum of states rather than a few sharp lines; no sharp tuning is required and a very wide range of resonant modes can interact with the electron population so that there is little control of the acoustic frequencies arising spontaneously. The same is true for the following examples which operate on similar principles. Frequency control is supplied by a resonant cavity and the devices may be tuned over a considerable range without losing the amplification from the non-resonant active medium.

Travelling-wave oscillator

A helix of wire, surrounded by a cylindrical metal sheath, as in fig. 7, acts as a coaxial transmission line on which electromagnetic waves travel rather

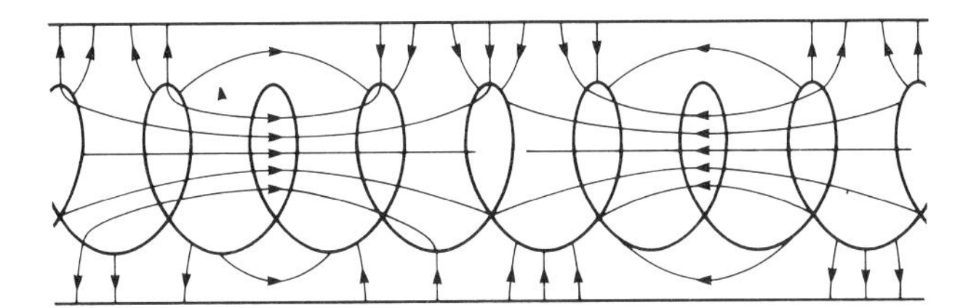

Fig. 7. Electric field distribution in a coaxial transmission line with helical inner conductor.

slowly on account of the inductance of the inner conductor. Moreover the electrical field-lines outside the helix terminate on surface charges which also generate a field component along the axis. This is one example of a wave which can interact with an electron stream moving faster than itself so as to be amplified. Unless we are to undertake a full analysis, such as is certainly desirable when the detailed design of an oscillator is the aim, what has already been said is enough to reveal the principle as virtually the same as for wave-amplification in CdS. Here, however, there is no problem arising from collisions (though space-charge forces may be significant) and the quantum theory of the interaction may be appreciated as having as good a status as the classical theory of bunching. Whereas in CdS the thermalized distribution of electron velocities spans a range much greater than the wave velocity, the reverse is true here; the electron gun can produce a rather well-defined stream of nearly mono-energetic electrons, so that by adjustment of the accelerating voltage the majority can be caused to move only slightly faster than the wave, and the regenerative interaction is correspondingly strong. In an actual device the amplified wave that emerges is fed back into the input so that the whole forms a ring resonator which may be imagined as spontaneously excited by the maser action of the inverted population of energy states in the electron beam.

The klystron

We shall devote more space to the klystron than to other oscillators, not because it is technically more important, but because in an idealized form its behaviour exemplifies both the bunching process and maser action with such simplicity that detailed analysis presents no difficulties. There are many variants of the design, according to the intended application, and we shall consider only one, the *reflex klystron* which for many years, until its supremacy was challenged by solid state devices, was the most convenient and reliable low-power continuous oscillator in the frequency range 10^9–10^{11} Hz. The resonator, shown in section in fig. 8(a), has axial symmetry and supports a fundamental mode which produces an oscillatory electric field in the gap at the centre. This region is in the vacuum tube (not shown) and an electron beam is accelerated from a gun at negative potential so that it passes through the gap and then is reflected back by the retarding field of the still more negative reflector electrode. It returns through the gap and subsequently is lost from the beam. We shall assume that the gap

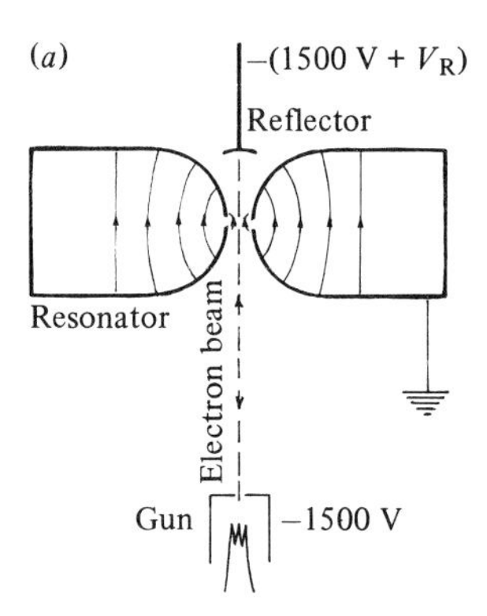

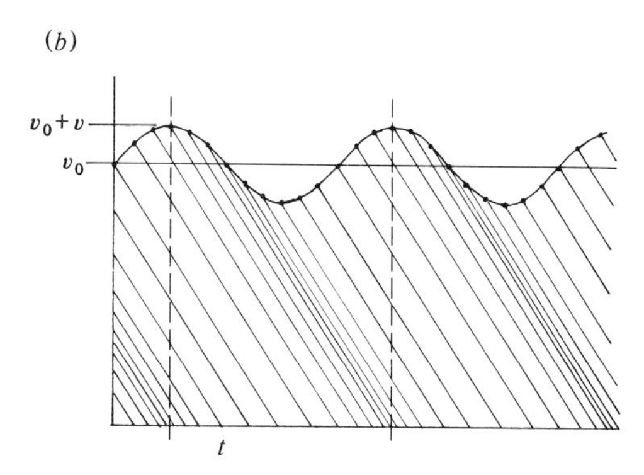

Fig. 8. (a) Schematic diagram of reflection klystron; the resonator, shown in section, is a torus with a gap round its inner surface through which the electric field spreads to interact with the electron beam. (b) Illustrating the bunching mechanism.

is very narrow and that the retarding field is constant between the resonator and the reflector. When the cavity is excited at frequency ω the electrons arriving at the gap with velocity v_0 leave with velocity $v_0 + v \cos \omega t$, being accelerated or retarded according to the phase of oscillation at the instant of passage. The same deceleration is then experienced by all, so that the fastest take longest to return, just like a ball thrown in the air. For values of ωt around $\pi/2$ successive electrons gain progressively less velocity from the gap and consequently return sooner so that the beam is locally more concentrated; half a cycle later the beam is correspondingly emaciated. A schematic representation is shown in fig. 8(b); an electron arriving at t returns to the gap after an interval $C(v_0 + v \cos \omega t)$, where $2/C$ is the deceleration. The lines, whose gradient is $-1/C$, show by their intersection with the horizontal axis the times of return for an originally equally spaced set of electrons, starting at values of ωt differing successively by $\pi/6$. It is clear that if the cavity excitation (i.e. v) is high enough, or the retarding field weak enough, the sinusoid may be steeper than the lines over part of its cycle, so that the electrons return with their order changed; where the straight lines are tangential to the sinusoid the local density on return is infinite (or would be if space charge did not prevent it). In principle, therefore, the bunching may lead to a highly non-sinusoidal distribution of charge in the beam, and indeed the klystron has proved a good device for frequency multiplication by virtue of its strong harmonic content.

Our concern, however, is with the self-excited oscillator, for which purpose we need only know the fundamental component of the bunching. This is determined by $\int_0^{2\pi/\omega} j(t')e^{i\omega t'}\,dt'$, where $j(t')$ is the current in the gap due to the reflected electrons. Now each electron, irrespective of its velocity, is responsible for a sharp impulse $j(t')$ such that $\int j(t')\,dt' = e$, its charge. To compute the Fourier integral, therefore, it is necessary only to sum $e^{i\omega t'}$ for each electron, with t' representing the instant of the return passage, and the difficulties that might have been anticipated from the peaked form of $j(t)$ are eliminated. Thus the electrons that made their first passage between t and $t + dt$ carry a total charge proportional to dt (we shall leave out all constants and look solely at the form of the behaviour), and arrive back when t' is $t+C(v_0+v\cos\omega t)$; it follows immediately that j_1, the fundamental component of the current, takes the form

$$j_1 \propto \int_0^{2\pi/\omega} e^{i\omega[t+C(v_0+v\cos\omega t)]}\,dt$$

$$\propto i\,e^{i\omega Cv_0} J_1(\omega Cv), \qquad (13)$$

where $J_1(x)$ is the Bessel function of the first kind and first order:

$$J_1(x)=\tfrac{1}{2}x-\tfrac{1}{16}x^3+\tfrac{1}{384}x^5-\cdots.$$

The meaning of this result is clear from fig. 8(*b*); $J_1(\omega Cv)$ is a measure of the degree of bunching, while ωCv_0 represents the phase of the bunches relative to the oscillating field. In the diagram the bunches have been drawn with their maximum density coinciding with the peak of the cavity field – the current maxima in the reflected beam meet a strong decelerating field and thereby transfer energy to the resonator. The ratio of j_1 to the field in the gap (to which v is proportional) may be regarded as the effective impedance presented by the electron beam when the behaviour of the resonant cavity is modelled by a series LCR circuit. In addition to the fixed circuit components there is an impedance of the form

$$Z_{\text{beam}}=-iA\,e^{i\omega Cv_0} J_1(\omega Cv)/v, \qquad (14)$$

where A is a constant determined by the beam density, gap configuration etc. In the figure $\omega Cv_0=3\pi/2$ and Z_{beam} is real and negative, providing the required regeneration. More generally R_{beam}, the resistive part of Z_{beam}; is equal to $A\sin(\omega Cv_0)J_1(\omega Cv)/v$ and the reactive part $X_{\text{beam}}=-iA\cos(\omega Cv_0)J_1(\omega Cv)/v$. If oscillation can be maintained, it settles down at an amplitude of v such that $R_{\text{beam}}+R=0$. Fig. 9 shows how the solution might appear under nearly optimal tuning conditions, $\omega Cv_0=3\pi/2$, and the beam is four times stronger than is needed to initiate oscillation. The

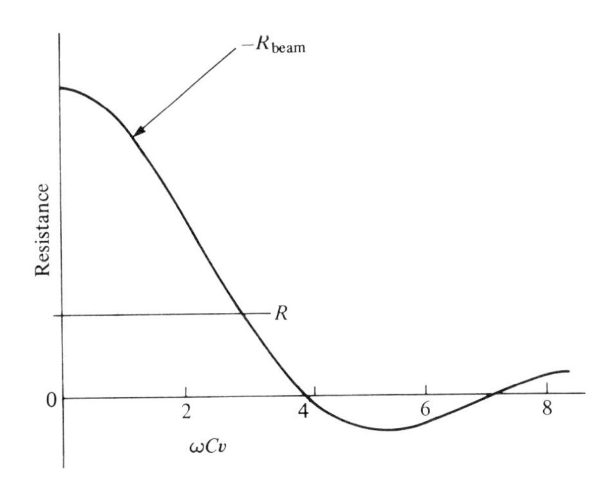

Fig. 9. Plot of the real part of (14).

amplitude builds up until v approaches the first zero of R_{beam}, which occurs when $\omega Cv = 3.83$, the first zero of $J_1(x)$. At this point $v \sim 0.8v_0$ and the beam is very heavily modulated. When there is, as in this example, plenty of amplification in hand, the reflector field may be varied over a considerable range, to reduce the magnitude of $\cos(\omega Cv_0)$, without stopping oscillation, and the consequent introduction of X_{beam} causes the frequency to shift. It is easily seen that the klystron is tunable by this means over a range several times the frequency band of the freely oscillating cavity.

Let us turn now to the quantum mechanics of the system, regarding the cavity field as a time-dependent perturbation of the electrons in the beam. This is a simple one-dimensional problem of a particle in a potential which, when the cavity is unexcited, is zero for $x < 0$ (between electron gun and gap) and equal to $2mx/C$ for $x > 0$ (between gap and reflector). Since the deBroglie wavelength is millions of times less than the dimensions, the WKB solution can be expected to provide a good approximation to the Airy function which is the solution for a linear potential variation, as in the region of positive x. For negative x a sinusoidal standing-wave solution applies, whose nodes are fixed by the need to join up with the Airy function. Now an electron of energy E has kinetic energy $E - 2mx/C$ at x, and wavenumber $(2mE - 4m^2x/C)^{\frac{1}{2}}/\hbar$; its phase change 2ϕ in going from gap to reflector and back is $2\int k\,\mathrm{d}x$, and may be written down immediately: **37**

$$2\phi = [2(2mE)^{\frac{1}{2}}/\hbar]\int_0^{x_0} (1 - x/x_0)^{\frac{1}{2}}\,\mathrm{d}x,$$

where $x_0 = CE/2m$ and marks the classical reflection point. Hence

$$\phi = \sqrt{2}CE^{\frac{3}{2}}/3\hbar m^{\frac{1}{2}}. \tag{15}$$

Every increment of E that increases ϕ by π shifts the standing wave at $x = 0$ from one node to the next. The unnormalized wave-function in the region $x < 0$ may therefore be written

$$\psi = \cos(kx - \phi)\,\mathrm{e}^{\lambda x}; \qquad x < 0. \tag{16}$$

The exponential $e^{\lambda x}$, where λ is very small, is added for convergence in the following calculation; it may be given physical plausibility by supposing that the electrons are injected into the beam all along its path towards the resonator, rather than at the gun, and are progressively scattered out after the return passage. This avoids having to invent a boundary condition for the gun that is physically realistic and allows no possibility of an electron making a second or third phase-coherent traverse of the beam.

The cavity in oscillation generates a perturbing potential $V' \cos \omega t$ in the form of a step-function, which is non-vanishing only in the range $x < 0$, where ψ takes the simple form (16):

$$V' = V_0; \qquad x > 0$$

$$= V_0 + V_1; \qquad x < 0,$$

where $V_1 = \frac{1}{2}m[(v_0+v)^2 - v_0^2]$. When V_1 is weak, the only case we consider, the strict periodicity of $V' \cos \omega t$ ensures that a given state E, having wavenumber k, is significantly perturbed only by the introduction of the states $E \pm \hbar\omega$. The probability of a transition occurring to one of these states, k' say, is proportional to the square of the matrix element:

$$M_{kk'} = \int_{-\infty}^{\infty} \psi_{k'}^* V' \psi_k \, dx = V_1 \int_{-\infty}^{0} \psi_{k'}^* \psi_k \, dx,$$

since orthogonality ensures that $V_0 \int_{-\infty}^{\infty} \psi_{k'}^* \psi_k \, dx = 0$. Hence, from (16), when $\lambda \to 0$,

$$M_{kk'} \approx \tfrac{1}{2} V_1 \sin(\phi' - \phi)/(k' - k), \tag{17}$$

the much smaller terms having $k + k'$ in their denominator being neglected. It is the difference in $|M_{kk'}|^2$ for upward and downward transitions that concern us. Binomial expansion of (15), with E replaced by $E \pm \hbar\omega$, shows that

$$\phi' = \phi(1 \pm \tfrac{3}{2}\varepsilon + \tfrac{3}{8}\varepsilon^2 \cdots), \qquad \text{where } \varepsilon = \hbar\omega/E$$

and the upper sign refers to the transition to $E + \hbar\omega$. Hence

$$\sin(\phi' - \phi) = \pm\sin(\tfrac{3}{2}\varepsilon\phi) + \tfrac{3}{8}\varepsilon^2\phi \cos(\tfrac{3}{2}\varepsilon\phi) + \cdots. \tag{18}$$

A similar expansion for k, which is proportional to $E^{\frac{1}{2}}$, shows that

$$1/(k' - k) = \pm 2/\varepsilon k + 1/2k + \cdots. \tag{19}$$

Hence

$$|M_{kk'}|^2 = (EV/m\omega v_0)^2[\sin^2\xi \pm \tfrac{1}{2}(\hbar\omega/E)\sin\xi(\sin\xi + \xi\cos\xi) + \cdots], \tag{20}$$

where $\xi = \frac{3}{2}\varepsilon\phi = \frac{1}{2}\omega C v_0$. Maser action is strongest when the negative sign gives the larger matrix element, i.e. when $\sin\xi(\sin\xi + \xi\cos\xi)$ takes its maximum negative value. The second term, proportional to $\xi \sin 2\xi$, gives the same condition as the classical calculation; if C is varied amplification is greatest when $C \sin(\omega C v_0)$ has its maximum negative value. The first term, which is relatively rather unimportant, is probably the result of inadequate approximations to the wave-functions. When the matrix elements are to be expanded to second order in ε the WKB approximation may require refinement; but we shall not attempt to achieve perfect agreement. At this degree of approximation the main point is clear enough, that bunching and maser action, as in the other examples, are alternative descriptions of a basic and essentially simple physical process of energy transfer.

Epilogue

As is appropriate in a book dealing with the quantum mechanics of vibrators, we started with a quantal calculation and have finished with one; but we have never strayed far from the classical models. Indeed the first chapter was largely concerned with establishing the validity of classical reasoning, and the very last problem of this chapter has been used to demonstrate, somewhat perversely it might be thought, that the method of quantum mechanics can occasionally be applied to systems which physicists and engineers would instinctively regard as classical. So long as the discussions give insight into physical processes and reveal the strengths and weaknesses of the analytical tools available, no apology is needed and no defence is offered except the evidence of the book itself. Even the rather simple systems which this volume and its predecessor have treated demonstrate clearly the increased power available to those who can handle both classical and quantal reasoning. The reader who wishes to apply his skill to complex vibratory processes will find his tasks eased if he can use whichever seems advantageous at each stage, and be confident that his understanding of simple processes will enable him to recognize the dangers inherent in both approaches – the danger of carrying classical reasoning too far into the quantum domain, and the danger of forcing over-simplifications on physical systems to make them amenable to the unforgiving methodology of quantum mechanics.

References

Chapter 13: The quantized harmonic vibrator and its classical features

1 J. L. Powell and B. Crasemann, *Quantum Mechanics*, p. 127 (Reading, Mass: Addison-Wesley, 1961).
2 L. Pauling and E. B. Wilson, *Introduction to Quantum Mechanics*, p. 67 (New York: McGraw-Hill, 1935).
3 Ref. 2, p. 81.
4 M. Abramowitz and I. A. Stegun, *Handbook of Mathematical Functions*, p. 475 (New York: Dover, 1965).
5 Ref. 1, p. 98.
6 G. M. Murphy, *Ordinary Differential Equations and Their Solutions*, p. 15 (Princeton, NJ: Van Nostrand, 1960).

Chapter 14: Anharmonic vibrators

1 A. Sommerfeld (tr. H. L. Brose), *Atomic Structure and Spectral Lines*, p. 232 (London: Methuen, 1923).
2 Ref. 13.4, p. 256.
3 Ref. 1, p. 304.
D. ter Haar, *Elements of Hamiltonian Mechanics*, 2nd Ed., p. 129 (Amsterdam: North-Holland, 1964).
4 Ref. 13.4.
5 J. C. P. Miller, *The Airy Integral*, (B.A. Mathematical Tables, Cambridge University Press, 1946).
6 Ref. 13.2, p. 271.
7 Ref. 13.1, p. 387.
8 Ref. 13.1, p. 140.
9 R. J. Le Roy, *Molecular Spectroscopy, Vol. 1*, Ch. 3 (London: The Chemical Society, 1973).

Chapter 15: Vibrations and cyclotron orbits in two dimensions

1 G. Herzberg, *Molecular Spectra and Molecular Structure, Vol. 2*, p. 215 (New York: Van Nostrand, 1945).
2 N. Minorsky, *Non-linear Oscillations*, p. 506 (Huntington, NY: Krieger 1974).
E. Breitenberger and R. D. Mueller, 1981, *J. Math. Phys.* **22**, 1196.
3 A. B. Pippard, 1964, *Phil. Trans. Roy. Soc.* A**256**, 317.
4 J. M. Ziman (ed.), *The Physics of Metals, 1. Electrons*, p. 118 (Cambridge University Press, 1969).
5 L. Onsager, 1952, *Phil. Mag.* **43**, 1006.
6 E. I. Blount, *Solid State Physics* (ed. Seitz and Turnbull), *Vol. 13*, p. 305 (New York: Academic Press, 1962).
7 P. J. Lin and L. M. Falicov, 1966, *Phys. Rev.*, **142**, 441.

8 Ref. 4, p. 145.
9 Ref. 4, p. 62.
10 Ref. 4, p. 129.
11 R. W. Stark, 1964, *Phys. Rev.*, **135**, A1698.
12 L. M. Falicov and H. Stachowiak, 1966, *Phys. Rev.*, **147**, 505.

Chapter 16: Dissipation, level broadening and radiation

1 B. I. Bleaney and B. Bleaney, *Electricity and Magnetism*, 3rd edition, p. 248 (Oxford University Press, 1976).
2 R. L. Sproull and W. A. Phillips, *Modern Physics*, 3rd edition, p. 653 (New York: Wiley, 1980).
3 Ref. 2, p. 258.
4 J. M. Jauch and F. Rohrlich, *The Theory of Photons and Electrons*, 2nd edition, p. 171 (New York: Springer-Verlag, 1976).
5 N. W. Ashcroft and N. D. Mermin, *Solid State Physics*, p. 519 (New York: Holt, Rinehart and Winston, 1976).

Chapter 17: The equivalent classical oscillator

1 D. H. Auston, Topics in Applied Physics (No. 18, Ultrashort Light Pulses, ed. S. L. Shapiro) p. 123 (Berlin: Springer-Verlag, 1977).
2 E. U. Cordon and G. H. Shortley, *The Theory of Atomic Spectra*, p. 133 (Cambridge University Press, 1935).

Chapter 18: The two-level system

1 Ref. 13.1, p. 358.
2 W. A. Phillips, 1970, *Proc. Roy. Soc.* A**319**, 565.
3 H. Fröhlich, *Theory of Dielectrics*, p. 70 (Oxford: Clarendon Press, 1949).
4 C. J. Gorter, *Paramagnetic Relaxation* (New York: Elsevier, 1947).
5 C. Zener, *Elasticity and Anelasticity of Metals* (Chicago University Press, 1948).
6 G. Frassati and J. le G. Gilchrist, 1977, *J. Phys. C.* **10**, L509.
7 H. W. Kroto, *Molecular Vibration Spectra*, Ch. 9 (London: Wiley, 1975).
8 Ref. 16.5, p. 425.
9 B. Bleaney and R. P. Penrose, 1947, *Proc. Roy. Soc.* A**189**, 358.
10 J. P. Gordon, H. Z. Zeiger and C. H. Townes, 1954, *Phys. Rev.*, **95**, 282.

Chapter 19: Line broadening

1 S. Ch'en and M. Takao, 1957, *Rev. Mod. Phys.*, **29**, 20.
H. Margenau and M. Lewis, 1959, *Rev. Mod. Phys.*, **31**, 569.
G. Peach, 1975, *Contemp. Phys.* **16**, 17.
1981, *Ad. in Phys.* **30**, 367.
2 D. Bohm, *Quantum Theory*, p. 611 (London: Constable, 1951).
3 J. Stark, 1914, *Ann. Physik*, **43**, 965.
4 B. Bleaney and R. P. Penrose, 1948, *Proc. Phys. Soc.*, **60**, 540.
5 T. W. Hänsch, M. D. Levenson and A. L. Schawlow, 1971, *Phys. Rev. Lett.*, **26**, 946.
G. W. Series, 1974, *Contemp. Phys.*, **15**, 49.
6 B. W. Petley and K. Morris, 1979, *Nature*, **279**, 141.
7 A. Abragam, *The Principles of Nuclear Magnetism*, p. 58 (Oxford: Clarendon Press, 1961).
8 B. Golding, M. v. Schickfus, S. Hunklinger and K. Dransfeld, 1979, *Phys. Rev. Lett.*, **43**, 1817.

Chapter 20: The ammonia maser

1 Ref. 18.9.
2 Ref. 13.4, p. 358.
3 M. O. Scully and W. E. Lamb, 1967, *Phys. Rev.*, **159**, 208.
4 Ref. 13.2, p. 79.
5 D. Kleppner, H. M. Goldenberg and N. F. Ramsey, 1962, *Phys. Rev.*, **126**, 603.
6 R. F. C. Vessot in *Radio Interferometry Techniques for Geodesy*, p. 203 (NASA conference publication 2115, 1980).
7 R. F. C. Vessot and H. E. Peters, 1962, *I.R.E. Trans.* (Instrumentation), 183.

Chapter 21: The family of masers; from laser to travelling-wave oscillator

1 A. H. Cook, *Celestial Masers*, Ch. 1 (Cambridge University Press, 1977).
2 D. J. Bradley, in *Ultrashort Light Pulses* (ed. Shapiro) p. 17 (Berlin: Springer-Verlag, 1977).
3 e.g. B. A. Lengyel, *Introduction to Laser Physics* (New York: Wiley, 1966), O. Svelto (tr. D. C. Hanna), *Principles of Lasers* (New York: Plenum, 1976).
4 A. H. W. Beck, *Thermionic Valves*, Chs 12, 14, 15 (Cambridge University Press, 1953).
5 I. B. Bott, 1965, *Phys. Lett.*, **14**, 293.
6 J. H. McFee in *Physical Acoustics* (ed. Mason), *Vol. IV A*, p. 1 (New York: Academic Press, 1966).
7 A. R. Hutson, J. H. McFee and D. L. White; 1961, *Phys. Rev. Lett.*, **7**, 236.
8 G. Weinreich, 1956, *Phys. Rev.*, **104**, 321.

Index